The Origin of the Aberdeen Angus Cattle
And its Development in Great Britain and America

by American Aberdeen Angus Breeders' Association

with an introduction by Jackson Chambers

This work contains material that was originally published in 1922.

This publication is within the Public Domain.

*This edition is reprinted for educational purposes
and in accordance with all applicable Federal Laws.*

Introduction Copyright 2017 by Jackson Chambers

Self Reliance Books

Get more historic titles on animal and stock breeding, gardening and old fashioned skills by visiting us at:

http://selfreliancebooks.blogspot.com/

Introduction

I am pleased to present another title in the "Cattle" series.

The work is in the Public Domain and is re-printed here in accordance with Federal Laws.

As with all reprinted books of this age that are intended to perfectly reproduce the original edition, considerable pains and effort had to be undertaken to correct fading and sometimes outright damage to existing proofs of this title. At times, this task is quite monumental, requiring an almost total "rebuilding" of some pages from digital proofs of multiple copies. Despite this, imperfections still sometimes exist in the final proof and may detract from the visual appearance of the text.

I hope you enjoy reading this book as much as I enjoyed making it available to readers again.

Jackson Chambers

E636.223
Am3a
cop.3

HIGHLIGHTS ON ABERDEEN-ANGUS HISTORY

1808—Hugh Watson bought six cows and heifers and a bull at the Tarnty market to start his herd at Keillor.
1811—"Black humble" exhibited from Aberdeenshire and won a prize at the Gairloch show.
1822—"Black cattle" classification at the Highland show in Scotland for first time.
1823—"Dodded animal" exhibited at the Highland show.
1824—Adam Bogue won grand championship over all breeds on an Aberdeenshire steer.
1829—Hugh Watson made first great exhibit of the "Black polls" at Perth.
1829—At Perth the breed was designated Angus.
1830—Tillyfour Herd founded.
1831—At Inverness it was called Aberdeen.
1834—At Aberdeen it was called Aberdeen and Angus.
1846—Old Jock appeared in show ring, "standing in the front rank of early sires to improve the breed."
1850—Duchess imported to America; first Aberdeen-Angus in New World.
1856—First foreign exhibition of the breed at Paris International Exhibition.
1862—International Fat Stock Show at Poissey, Aberdeen-Angus won four of six prizes for "polled oxen."
1862—Vol. 1 of herd book issued in Scotland.
1867—Black Prince carried all before him, crowning his career with victory at Smithfield show, London.
1873—First bulls of breed brought to America by Thomas Clark and George Grant and taken to their Kansas ranch to improve the range cattle.
1876—Prof. Brown of Ontario Agricultural College imported first breeding herd for America.
1878—Universal Exposition, Paris, gave Aberdeen-Angus grand championship for group of foreign bred cattle and supreme championship for best beef-producing animals on the grounds.
1878—Findlay & Anderson, Lake Forest, Ill., imported first pure-bred Aberdeen-Angus herd to the United States.
1880—First steers carrying Aberdeen-Angus blood came to market in America, Jos. H. Rea & Son, Carrollton, Mo., feeding and marketing them at Chicago.
1880—Polled Cattle Society formed in Scotland.
1881—Smithfield "age limit" for grand champions lowered when an Aberdeen-Angus heifer won the honor.
1882—Championship for best cow at Kansas City show won by an Aberdeen-Angus.
1883—American Aberdeen-Angus Breeders' Association organized.
1883—Over 800 head imported from old country.
1883—Black Prince, steer, shipped from Liverpool for show at Kansas City and Chicago.
1884—G. W. Henry showed "best cows" at Kansas City, also winning slaughter test.
1885—James J. Hill showed Benholm at Chicago; dressed 71.4%.
1886—X. I. T. Ranch began a great test of the three breeds on the range.
1887—Turlington herd won grand championship at Chicago with Sandy.
1893—World's Columbian Exposition, Chicago, won supreme championship for the breed.
1900—International Live Stock Exposition inaugurated at Chicago.
1900—Dispersion of Estill & Elliott herd, Estill, Mo.; 58 females averaging $583, and 14 bulls $561.
1902—Record Association headquarters moved to Chicago.
1904—Aberdeen-Angus won all four inter-breed steer grand championships at the International.
1919—THE ABERDEEN-ANGUS JOURNAL founded at Webster City, Iowa.
1919—Enlate sold in public sale at $36,000.

FOREWORD

During the past decade a nation-wide interest in Aberdeen-Angus cattle and consequent increase in the number of breeders and members of the American Aberdeen-Angus Breeders' Association has created a new demand for information concerning the breed. To meet this call—to place before all seeking it information concerning the greatest and oldest beef breed—this short history of its origin and development in Scotland, its transplantation to America and its subsequent career has been prepared. No attempt has been made to present more than the merest outline of the breed's history, but the main facts have been brought down to date. Controversies long and bitter have prevailed touching more than one phase of the breed's development. No cognizance has been taken of them, the generally accepted view having been given in every instance. Differences as to men and matters incident to the breed's progress 50 to 100 years ago can have no possible bearing on its position today. The plain, unvarnished truth has been told, with a fair field for all and favor toward none.

For an extended and diverse history of the Aberdeen-Angus breed, from which much of the matter in this little work was taken, the Association desires to express its thanks to Mr. R. C. Auld, New York, a nephew of the late Wm. McCombie of Tillyfour. Mr. Auld's close relational and territorial connection with Mr. McCombie and his possession of many of the great breeder's private records and memoranda, fit him eminently to give potent aid in the preparation of such a history.

From this short history and from the other statistical and descriptive publications of the American Aberdeen-Angus Breeders' Association a complete review of the Aberdeen-Angus breed's history may be obtained from its Caledonian birthplace to the proud pinnacle of fame on which it now rests wherever good beef is grown.

CHAS. GRAY,
Secretary.

Chicago, May, 1922.

THE LINEAGE OF THE BREED

Hornless cattle existed in the earliest age to which we can trace the bovine form. In the day when first humanity itself appeared—when man was savage and a hunter, but yet an artist—there existed a finely formed polled race of cattle as depicted on the walls of the caverns of these dark-skinned folk—"the oldest of men." During that period Britain was continuous with the continent and consequently the polled cattle spread into the nooks and corners of what was then the land's end of Europe. Fossil and semi-fossil remains, found in Scotland, establish the fact of the prehistoric existence of wild polled cattle in those very districts occupied by the known ancestors of the present Aberdeen-Angus breed. King Kenneth MacAlpine of Scotland when promulgating the laws at Scone in Perthshire, specifically mentions "black *homyl" cattle—which is the first historical reference to the breed we have.

Memorial stones erected by the native Scots in commemoration of their repulses of invading Norsemen, and now found at Aldbar and Meigle in Forfarshire and Burghead in Morayshire certainly depict "the hornless cattle of the country." Kenneth's laws applied to the region that became the early seat of the Aberdeen-Angus breed, and there is documentary evidence to show that in 1523 the black homyl cattle occurred in Aberdeenshire, but the first specific mention of polled animals of the breed is that of the prize-winning "black humble" exhibited in 1811 from a well-known Aberdeenshire source, the influence of which on the breed is still felt. These concentrated facts prove that the Aberdeen-Angus is the oldest polled breed—in Britain at least—and accounts for its extraordinary prepotency in transmitting its color and hornless character. The value of these characteristics in crossing or grading up—in putting the royal stamp of the market-topping supremacy on the progeny of otherwise untrademarked stock—is, in the light of these marshalled facts, as easily explained as it is unchallenged.

THE BREED'S BIRTHPLACE

Scotland's earliest historians describe the region comprising the Northeast counties as a famous grazing ground for cattle and point out that Buchan was so called because it paid its tribute to the Roman legionaries in cattle. Church and state encouraged the improvement of the native stock, the Church being responsible for the establishment of the great Scotch fairs or trysts, which were originally gatherings held on days dedicated to the saints. Hence at a very early date they had the Aikey, Paldy and Tarnty Fairs in Aberdeenshire, Kincardineshire and Angusshire, and these fairs, as the evidence proves, were the first places where the breeders disposed of their surplus stock, which was eagerly snapped up by the English graziers. With the great increase in the demand for beef, as a result of the Queen Ann wars

---

*"Homyl," "humble," "humlie," "doddie" and "dodded" are all Scotch equivalents of the words "polled" or "hornless."

(1664-1714), cattle breeding in Scotland received a great impetus. Englishmen—always great beef eaters—found that their choicest meats came from the Scottish side of the border and it was while this trade was at its height—in the middle and latter portions of the 18th century—that authentic record was made of the breed's progress in Aberdeenshire and Angusshire, Kincardineshire and Morayshire. In Aberdeenshire we find the breed described as having been "improved" by putting the best males to the best females, rules being laid down for breeding both for beef and the dairy, which proves the breed to have been originally what we term a dual-purpose one.

The breeds of Aberdeenshire are fully described and innumerable descriptions are given of the fine specimens killed in the City of Aberdeen. The butchers or fleshers there had their own guild and were men of a high class. They were closely associated with the breeders and the dealers—and among these particular mention is made of the Williamsons of St. John's Wells and Robert Walker of Wester Fintray. Both these were breeders—though the Williamsons were also the largest dealers in Scotland, their only rival being, as we are told, Charles McCombie of Tillyfour, father of William McCombie, who thus inherited his love of the native humlies—as the breed was called—the word being the same as the homyl of Kenneth's laws, and of Aberdeenshire in 1523. The blood of the Williamson and Walker cattle made a distinct mark in the early showyards of the Highland Society. Robert Walker of Wester Fintray had in fact exhibited "black humble" cattle at the Gairloch Show, in Aberdeenshire, in the year of 1811, and Charles McCombie of Tillyfour judged them. The first mention of a "dodded animal" in connection with the Highland Show was of an Aberdeenshire polled steer, exhibited in 1823.

In the meantime operations had been also without doubt going on in Angusshire, especially in the eastern part of the county— the Brechin district, the great Fair of which was Tarnty. There the "First Families" of the breed are heard of—the Keillor Jocks and Favorites, the Buchan Black Megs and Panmure—to which sources may be traced the great improvement that then appeared in the breed. It was to the great Tarnty market that Hugh Watson resorted in 1808 to buy six cows and heifers and a bull, which were derived from farms in this section and Kincardineshire, with which to start his herd at Keillor. Old Grannie, the Prima cow, is believed to have been among the lot, and purchased originally from a Kincardineshire breeder. From here also the first of the Jocks came.

Among the other early breeders of this formative period were, besides Fullerton of Ardovie and the others indicated, Mustard of Leuchland, Ruxton of Fannell, Scott of Balwyllo, Dalgairns of Balgavies, Chalmers of Aldbar, Kinnaird (Lord Southesk) and Bowie, Mains of Kelly, the great bull breeder.

In Kincardineshire the early breeders of the formative class included Hector of Fernyflatt (of "sire of Panmure" fame); Silver of Netherly, and Black Meg, the dam of Panmure fame; Sir Thos.

and Alex. Burnett of the Leys; Scott of Easter Tulloch; McInroy of The Burn, and Portlethen. Portlethen is the oldest herd today—having had a continuous existence from 1818, which might be extended beyond that, if we take into account the herd of the previous occupier, Mr. Williamson, breeder of the polled Aberdeen bull, Colonel. But 1818 is enough for the purpose.

In Aberdeenshire, the list included the Williamsons and the Walkers, unique in these animals; the Earl of Kintore; McCombie of Tillyfour, Pirrie of Collithie, Wilson of Netherton of Clatt, Col. Fraser of Castle Fraser, whose herd was a famous show herd; Rev. Mr. Brown of Coull, Lamond of Pitmurchie, Conglass of Culsh, Walker of Ardhuncart, whose herd was founded about 1812 by purchases from Wester Fintray; Cooper of Hillbrae, of Earl o'Buchan fame; besides the breeders and exhibitors of these prime specimens of the breed that appeared at the early Aberdeen shows.

In Banff was the herd of Sir John Macpherson Grant, from which is descended the present premier herd as it exists today; Brown of Westertown; Walker of Montbletton (with close Fintray affinities); Collie of Ardgay; Skinner of Drumin, and Paterson of Mulben. In Morayshire were Brown of Linkwood and others.

Besides these there were breeders in Cromarty and Rossshire, as the annals of the Highland Society's show prove—all breeding "Aberdeens."

At the earliest shows of the Highland Society, before the first regular appearance of animals of the breed—which occurred in 1829—we have proofs of its prominence. In 1822 there was a general classification of "Black Cattle," which included both polled and horned. At the show of 1823, at the same place, an Aberdeenshire "dodded" steer is mentioned. In 1824 at Edinburgh, we find Adam Bogue exhibiting an Aberdeenshire steer that took the prize open to all breeds, weighing "1,225 lbs., sinking the offals," or 2,145 lbs. live weight. At the 1825 Edinburgh show, an Aberdeenshire steer of enormous size was also exhibited, selling for $225. In 1827, the last of the first series of shows, Mr. Bogue obtained the second prize for a steer bought from, this time, "Mr. McCombie of Tillyfour, at the Falkirk 1826 September Tryst—this being William, not Charles, McCombie.

HUGH WATSON OF KEILLOR

Hugh Watson, whose father and grandfather, like McCombie's, had been admirers of the native breed, began his herd in 1808, and in that year he proceeded to Tarnty to buy the best six heifers and a bull he could get. These animals were derived from West Scryne, Kinnaird, and Fannell in Eastern Forfarshire; and, as is also believed, some of them from Buchan—Old Grannie, understood to have come from Kincardineshire, dropping a calf—her first, at Keillor—named Beauty of Buchan. Alex. Bowie, Mains of Kelly, began to breed in 1809—a year after Mr. Watson; and William Fullerton of Ardovie was contemporaneous with

Tillyfour—the latter beginning to breed in 1830, and Fullerton about the same date—it being in 1831 that he bought the Black Meg, which "ranks with the Prima cow" and from whom the "Queen's breed" at Tillyfour was to spring. It is in the herds of these four breeders that we must seek the blood, and study the operations conducted therein, to understand how the breed started on its career. Little is known of Mr. Watson's breeding operations. For instance, it is not generally known that he had a sale in 1818, at which he offered animals of the "improved dodded breed."

Hugh Watson made the first great exhibit of the black polls at Perth Show in 1829. With one exception all the animals in the class were from the Keillor herd, and with one exception—a cow purchased by him from Peter Watson, Kirriemuir, "a dealer in Aberdeenshire cattle"—all were also bred by him. No names are given of the animals. Mr. Watson also showed the only Angus steers exhibited, one pair bred by himself, the other by Mr. Johnstone, The Scryne. The admiration for Mr. Watson's bull, cow and oxen was necessarily universal, but "it created surprise that in such a county the shrinking from competition should be so complete." Mr. Watson also exhibited at Kelso in 1832; Aberdeen in 1834; Perth, 1836; Dundee, 1843; Inverness, 1846, and lastly at Perth, 1852.

AT INVERNESS AND ABERDEEN

At Inverness, 1831, a very interesting exhibit appeared. The first prize aged bull was exhibited by Peter Brown, Linkwood, Elgin, who was mentioned by Wight at the end of the previous century as breeding "the best of the country breed." At this show also appeared perhaps the first cross of the Shorthorn and the Aberdeenshire black poll—a bull exhibited by none other than the great Barclay of Ury, Kincardineshire.

As Inverness had been, like Perth in 1829, the first territorial opportunity for the breed, the class being for the Aberdeen Polled, to appear before a national public, doing so in a distinctly impressive manner, so we find it natural that coming to Aberdeen right in the home of the old breed, a stunning exhibit of Aberdeens should appear. But there were also Angus, three bulls, two of them from Keillor. This was the first time Keillor met the Aberdeens in the showyard, and the first prize went to the Aberdeenshire polled bull exhibited by Mr. Findlay, Balmain.

The first prize for cows at the Aberdeen show was awarded to Lord Kintore; and the second would have gone to him also, had the regulations permitted. Both were bred at Wester Fintray. Strangely, the cow that was given the second prize was exhibited by another Walker still—Robert of Portlethen. There were 17 competitors in the class. In the heifer class Mr. Hector of Fernyflatt, the breeder of "the sire of Panmure," was first. The "great feature of the show was the exhibit of the polled." Here the class was for "Aberdeen and Angus" cattle—the two previous shows having doubtless demonstrated the affinity of the two "breeds." At Aberdeen there were prizes for dairy produce, and Mr. Walker, Wester Fintray, came out strong in the winnings—as noted in the chapter on "Milking Properties."

At Perth in 1836, Mr. Watson was again alone, winning in both sections. At Inverness, 1839, Keillor was first in the aged bull class for a bull bred at Balnacreich; and Ferguson of Pitfour, Aberdeenshire, was first in the junior class. Among the cows the Duke of Richmond's took first to Wester Fintray's second. The Duke exhibited a lot of ten polled cows, three having produced twins. Fine polled Aberdeenshire steers were shown, and high prices obtained for them.

At Aberdeen, 1840, there was a great show. A Kincardineshire poll headed the bulls—his portrait being published in the Farmer's Magazine, where he is described as an "Aberdeenshire Polled Bull." Among the cows, we find Wellhouse exhibiting and Portlethen; but Wester Fintray was too strong for them, being first for cows; Dingwall of Bruckley was second; Duke of Richmond third. In the two-year-olds the Duke was first and Walker second. In yearlings, Portlethen was first.

The show at Dundee, 1843, saw the clash of battle between the rival camps—all Angus was out in force—Keillor also exhibiting Shorthorns; but he had to bow to the irresistible presence of the Panmure and Black Meg host. The blood of these contingents was conspicuous and drove competition before them—competition that had ruled Keillor, Portlethen and even Wester Fintray. Lord Panmure also exhibited oxen. Tillyfour was likewise an exhibitor of oxen. It was soon after this that Panmure, "the Hubback of the Polls," went north to the Wellhouse herd; and Queen and others of her family went to Tillyfour—all from William Fullerton's sale.

Again at Inverness, 1846, Mr. Watson was first with Old Jock. Note that in 1843 and 1846 appeared Panmure and Old Jock said by the best authorities "To stand in the front rank among the early sires that have most contributed to the improvement of the breed."

TILLYFOUR AND ITS INFLUENCE ON THE BREED

At some point in the record of every improved breed we reach a man who possessed a peculiar genius for committing his impressions and an account of his operations to writing. "When we come to deal with Mr. McCombie, we stand on firm ground," it has been written, and in private herd records and other personal memoranda of Wm. McCombie of Tillyfour a safe historical base is found. The McCombies were an "Aberdeen and Angus" family of ancient derivation and for many years famed as cattlemen. The Tillyfour herd was founded in 1830 with animals accredited by Mr. McCombie himself as "Aberdeens," as from St. John's Wells and Wester Fintray, and as "Angus," from Keillor, Balwyllo, Dalgairns and elsewhere.

We find here in this assembling of animals of these two districts, which took place previous to the first territorial Highland

Show at Aberdeen 1834, both "strains" of the breed; in which assembling and matching we also witness the welding of the breeds, to which the designation was given for the first time, on that account, of "Aberdeen and Angus;" a title that years after its fortunate consummation was recognized by The Polled Cattle Society, which was constrained to follow the approved mode adopted by the American Aberdeen-Angus Breeders' Association; the modern title remaining that of Tillyfour with the simple replacement of the connecting "and" by a connecting hyphen.

At Perth, 1829, the breed was designated Angus; at Inverness, 1831, Aberdeen; at Aberdeen, 1834, Aberdeen and Angus, some specimens being actually from Forfarshire. The actual welding of the breeds, however, took place when specimens of both were mated together knowingly—as at Tillyfour—and at a much earlier date with the Black Megs and other famous specimens that found their way into Eastern Forfarshire.

"THREE FOUNTAIN HEADS OF THE BREED"

So, now, we may take up the task of briefly pointing out the great families of the breed, that have been so well cultivated.

In "The History of the Breed," we read, referring to Dr. Thomas F. Jamieson's researches into the foundation history of the breed, (quoted from him): "When I occupied the post of Fordyce Lecturer at Marischal College, Aberdeen, I devoted some attention to the subject of polled cattle along with other matters, and I found that all the best blood of the Aberdeen and Angus doddies trace back to three fountain-heads, viz: 1st, Mr. Fullerton's Black Meg; 2nd, the bull Panmure, from Brechin Castle, and 3rd, the Keillor Jocks." A well known writer, referring to the former says, "Black Meg, the mother of Queen, ranks with the Prima cow" as a foundress of the breed. She has also been likened to the "Favorite cow among the Shorthorns." Panmure, also, has been termed "the Hubback of the Polls." His dam was also called Black Meg—and there used to be some confusion between the two which now does not occur—so familiar have they both become. "Mr. Fullerton's Black Meg," which that breeder secured about 1831, was the foundation cow at Ardovie, where she produced Queen of Ardovie, the foundress of the Tillyfour Queens and Prides. A daughter of Queen and granddaughter of Black Meg, viz: Princess (831) was purchased by Mr. Watson of Keillor and was sent to Ireland to uphold the standard of Keillor there. She dropped a heifer calf at Keillor. Her calf, by Adam, was secured by Mr. Ruxton, Fannell.

The other Black Meg—dam of Panmure—calved 1837, was bought for Lord Panmure in a lot of eight or ten heifers from Mr. Silver of Netherley, Kincardineshire—mentioned in the "General View" of that county, and were known to have been bought in Buchan originally, passed from Panmure eventually to The Scryne, whence had come some of Keillor's Tarnty selections. Mr. Fullerton describes these Black Megs as Beautys. Black Meg and Beauty, indeed, seem to have been favorite names for these

Buchan cows. The genial Hugh Watson named the calf of Old Grannie—herself remembered as one of the Tarntys that also came from Kincardineshire—"Beauty of Buchan"—a very significant designation, reminiscent most probably of the origin of the calf's dam herself. Mr. Watson also named another heifer by the magic name "Panmure" at Keillor, which shows he had his mind on the sort.

Black Meg of Panmure was bought by Mr. Bowie of Boysack, brother of Alexander Bowie, Mains of Kelly, who had his eye upon her, and was able to gain possession of her. In his herd she founded the famous Martha and Mary families. Major, a noted sire at Mains of Kelly, was of the Martha family, siring Gainsborough, the first prize bull at Inverness, 1874, Gainsborough's son, Logie the Laird, being first at Edinburgh, 1877.

Thus, we find these two Black Megs' progeny going into the four foundation herds of the breed—Ardovie, Mains of Kelly, Keillor and Tillyfour. In a letter written by William Fullerton, to the writer, dated in the spring of 1879, he said that he purchased Black Meg (dam of Queen) from Mr. Thos. Fawns, a well known cattle dealer in Brechin (mentioned by Mr. McCombie in "Cattle and Cattle Breeders," as a frequent companion) about the year 1853. She was calved in 1831 and he wrote enthusiastically about her, naming her over and over again as a "pure Buchan doddie." These Black Megs represented the sort of stock that was reared at St. John's Wells, Wester Fintray and all the other original seats of the breed in Aberdeenshire from which the Tillyfour herd was collected; and also of the kind that appeared at Inverness, 1831, and at Aberdeen, 1844, and other early shows.

Mr. Fullerton in the communication referred to said: "Give me ten good old-fashioned cows of the old Buchan kind, and let me put an Angus bull, with the best head, and neck which can be found, to them, and I'll venture the offspring will all take prizes."

Old Jock, who took the first prize at Dundee in 1843, where Panmure stood first proudly in the senior ranks, is the Jock referred to by Dr. Jamieson, the sire of most of Mr. Watson's stock from 1843 to 1852. His dam was Old Favorite. He was the sire of Angus (45) that aided in the "welding of the breed." He was also the sire of Emily (332) the foundress of the Ballindalloch Erica family. Old Jock was likewise sire of Emily of Kinochtry, thus founding an extension of the Old Grannie foundation.

Thus we are enabled from these sources to specify the several most famous families of the breed.

First, in a word, the Tillyfour Queens (Queen Mothers) and Prides—of many branches and ramifications.

Second, the Ballindalloch Ericas, through Eisa and Enchantress, which need no words to describe them.

They follow the Ballindalloch Jilts and Tillyfour Ruths from Beauty of Tillyfour, bred at Keillor from which she was bought by Mr. McCombie in 1860 for $320. This Beauty was also the foundress of the Easter Skene Miss Watsons. Then come the Kinochtry Princesses, Emilys (mentioned) and Favorites descended from Old Grannie and Favorite.

The Kinnaird Fanny family descends from Old Lady Ann (743), calved four years before Old Grannie, and possibly the oldest cow in the Herd Book. Among the Portlethen families the Mayflower represents one of the oldest strains, descending from Old Maggie (681). But the most interesting family of all associated with this, the oldest herd in existence, is the Nightingale (262) family—derived from Mary of Wester Fintray (21). The Montbletton Mayflowers and Lady Idas have a grand reputation. Here the foundation blood traces back also to Wester Fintray.

The Mains of Kelly Marthas and Marys—Black Meg families—have been mentioned; which sort of blood seemed to suit Mr. Bowie, for he got hold of Young Jenny Lind from Tillyfour with which he founded his Jennets. The Drumin Lucy family is one of the old sort. Castle Fraser produced the Blanche tribe from which Geo. Bruce's first prize cow at the Paris Exposition in 1878 was derived. All the Castle Fraser blood, wherever found, may be set down as of the finest sort. Westertown evolved the Rose family from Blinkbonny (315).

Then we have the Mulben Mayflowers; the Rose of Advies, the Ardgay Zaras; the Fyvie Flower family—which should have a touch of the old St. John's Wells influence. The Abbesses and Actresses of Easter Tulloch created at Balquhain from which have sprung some wonderful champions in Britain and America. The Cortachy Ariadnes maintain an interesting tradition. Old Jip (965) founded the Jipsey family at Spott, where Mr. Whyte maintained an old herd.

At The Burn, Col. McInroy established the Matilda of Airles that have furnished champions at the Highland and Royal shows.

A comparatively modern family, but one which was derived from the best sort of foundation, is the Aboyne family of Saints. The foundress of the family was Sibylla, a finely formed and superbly fleshed cow. The family produced champions one after another, at the Highland and the Royal Northern shows. Of the same class are the Waterside Matildas, which showed the value of the Pride top on such good old-fashioned stock as the Indigo, Tarland herd possessed, Matilda being by Knight of Shire, whose career as a stock bull was of very limited duration, indicating, however, that if he had lived he would have made a record as a sire that would have been unique.

Among sires that made their mark in the earlier days of the breed were Old Jock and Angus Young Jock; Panmure, Monarch and Victor, Earl o'Buhcan, Hanton, Cupbearer, Druid, Palmerston, Windsor, Rox Maule, President of Westertown, Clansman, Trojan, Justice and Young Viscount, Gainsborough, Duke of Perth, Shah, Prince Albert of Baads, Prince of the Realm, Knight of the Shire, The Black Knight, Black Rod, Cash,

Epigram, Prince Inca, which have been followed by a host of bulls of the highest merit, showing the steady gain in quality of the breed.

Notable sires of the breed in the Old World, as recorded by Secretary James R. Barclay, of the Aberdeen-Angus Cattle Society, Aberdeen, Scotland, include Jim of Morlich, Erica Prince E., Loyalist of Morlich, Eblito by Prince Ito, Lord Sirdar of Advie, Just Rover of Morlich by Rover of Craibstone, Elate by Mailbag, Premier of Finlarig, Delamere, Postman of Aberlour, Eryx of Ballindalloch by Prince of Insh, Just Rover of Morlich 2d, Juba of Morlich by Rover of Craibstone (Juba was first at the Highland and was exported to the United States in 1902 to head the late John Campbell's herd in Minnesota, now Campbell Brothers), Prince Ito, Jeshurun by Eblito, Evictor by Eblito, Evarra by Mondamin, Wizard of Maisemore by Elate, Khartoum of Ballindalloch by Delamere (Khartoum was imported by C. Herendeen, to head his Illinois herd), Ebbero by Delamere, Edelhof by Eblito, Publican of Preston, Eblamere by Delamere (Eblamere sired the Perth champion Earl Eric of Ballindalloch that was imported by Escher & Ryan to head their herd and make show and sale ring history with his get), Elector of Benton by Esmond of Ballindalloch, Prince of Ake, Bion, Benin of Ballindalloch by Eblamere, Everwise by Wizard of Maisemore, Gabriel of Eshott by Idelamere, Escape of Towiemore, Evolen of Ballindalloch by Eblamere, Euthalito by Eblito, Erops by Echevin, Evard of Ballindalloch by Eblito, Prince of the Wassail by Delamere, Nicotin, Eudoxus by Juba of Ballindalloch, Edensor by Delamere.

Mr. Barclay quotes an Old Country authority, probably the late Mr. John MacPherson of Mulben, in naming Iliad, Delamere, Prince of the Wassail, and Baron Beauford as the four most influential bulls of the breed in the last half century. The latter bull is still living and heads the Bleaton herd. These four are descended in a straight line of breeding, Baron Beauford being by Prince of the Wassail, he by Delamere, and the latter being a grandson of Iliad.

Jason of Ballindalloch at the head of the Harviestoun herd, Eureka of Advie in the Skillymarno herd, Evilesco and Elorus of Ballindalloch in the Ballindalloch herd, are living bulls on the other side that have already left their stamp on the breed with sufficient impress to be noted throughout the Aberdeen-Angus world.

Imported bulls on this side have been scattered thinly, widely and sparingly in the past fifteen or twenty years. Since the days of Prince Ito, Bion, and Eliminator of Ballindalloch, three of note on the other side before importation to the United States we have had only a scattering few which have not been placed in the hands of the men making breed history on this side with very few exceptions. Edgar of Dalmeny in the Wildwood herd in Michigan, Evenest of Bleaton in C. W. Eckhardt's New York herd, Perinthian in Dr. J. I. Huggins' Tennessee herd, Proud Eric of Aberlour in Mr. Peabody's California herd, Just Eric of Aberlour in

Mr. Harrison's North Carolina herd, Edgardo of Dalmeny and Elcho of Harviestoun in the Woodcote herd of Michigan, and Eclipse 2nd of Morlich in the Campbells' Minnesota herd have all yet to make their mark in the breeding annals on this side, though the New York and Tennessee bulls have already gained grand championships at the Chicago International to stamp them as show bulls of the highest merit.

The International Live Stock Exposition show records indicate that the following bulls have impressed themselves on the breed in the United States sufficiently to gain recognition: Heather Lad of Emerson 2d, Black Woodlawn, 10th Laird of Estill, Baden Lad, Earl Marshall, Heather Blackbird, Black Monarch of Emerson, Rosegay, Bonnie Ben Royal, Undulata Blackcap Ito 2d, Oakville Quiet Lad by Black Woodlawn, Ensign of Glencarnock, Idolmere by Oakville Quiet Lad. Following this list, much argument would undoubtedly arise as to bulls that should be named for a longer list. These, however, have left sufficient impress to go undisputed in the last twenty years, Baden Lad and Earl Marshall especially impressing their names with four International Live Stock Expositions' "get" victories each.

OTHER GREAT HERDS IN SCOTLAND

Among those that followed the foundation herds we may mention first the Aboyne Castle herd, of the Marquis of Huntley who has been a constant patron of the breed, was the first president of the Polled Cattle Society, and a noted exhibitor. His Lordship, especially in his last herd, showed his good judgment in selecting and breeding the Saints, the foundress he obtained from the Bridgend herd. We might mention Duff House where we find Lilas of Tillyfour and Young Viscount and other splendid specimens; Easter Tulloch, whence the Witches of Endor of Paris fame were originally derived; the Altyre herd of Sir William Gordon Cumming, that produced the Smithfield champions, must always be remembered with gratitude; Fyvie Castle and "Sir Maurice" and the herd of Mr. George Reid. The Baads were especially noteworthy. Castle Fraser we have mentioned; and Montbletton, Westertown and Mulben. Mr. Smith's herd at Powrie produced many attractive animals. Rothiemay was long noted and Kate of Glenbarry is a pleasant memory. Guisachan was a splendid herd full of choicest strains; Haughton came to the front wonderfully towards the latter part of its existence. The Johnstone herd, Waterside and Wellhouse, were strong in good blood—the former's Waterside Matildas showing the impress of Knight of the Shire.

The great herds at present include the Royal herd at Abergeldie Mains, founded by Her Majesty, the late Queen Victoria, who took a great interest in the breed. The herd, indeed, was formed soon after the visit was paid by Her Majesty to Tillyfour in 1868. It is strong in Prides and Saints, Ericas and Georginas. The herd has won the Family Prize at the Royal Northern and the female championship at the Highland Show of 1896.

The Auchorachan herd of Col. George Smith Grant was one

of the strongest in Ericas and Prides and old-fashioned strains. Record-making prices have been secured at its sales. It has been wonderfully successful in breeding champion showyard animals—we all remember Young Bellona, Clement Stephenson's Smithfield champion of 1887. It certainly has had a spectacular history. It is still continued in the Mains of Advie herd.

The herd of Alexander and James Beddie, Banks, Strichen, produced two of Stephenson's Smithfield champions, those of 1885 and 1894—Bridesmaid of Benton and Benton Bride, respectively.

The Collithie herd is still flourishing. The Conglass herd has certainly a record that is hard to beat—in the way of producing the sort of animals that advertises the breed in the best way—those prize steers that attract so much attention at Birmingham and Smithfield. Mr. Stephen supplied Mr. McCombie with a steer that won alongside the ox that captured the Prince Albert Cup at Poissy in 1862—and he was "heavier than Black Prince" of 1867. The Cortachy herd, of the Airlie family, progresses along the best lines; the former Earl having given the herd his personal attention, he and the late Earl of Southesk finding in each other a congenial subject that led them to work in harmony.

Mr. William Wilson at Coynachie and Drumfergue maintains perhaps the largest herd in Scotland—close to 200 animals. His Queen Mother of Drumfergue 7th was imported to this country where she had a noted showyard career and was sold for $3,000.

The Cullen House herd of the Countess of Seafield (now dispersed), was noted as another breeding herd that has produced a Birmingham and Smithfield champion, in 1908, from the family styled Her Majesty of Cullen.

The Dalmeny herd of the Earl of Rosebery has been a noted producer of Smithfield form—female champions. Its sire, Ebbero, the highest priced bull sold by public auction in Scotland up to the war, was the sire of Dalmeny Lady Ida 3d, the female champion at Smithfield, 1906. The herd has won the female championship three times at Smithfield and other equally high honors. Mr. Andrew Mackenzie of Balmore has a magnificent herd to which he is devoting his skill, in building up the Lady Ida family, an offshot of the Montbletton herd.

In Ayrshire, Mr. Kennedy of Doonholm has a choice collection of doddies. It has been extraordinarily successful in the showyard especially with females and in the export and sale department.

At Glamis there is a show herd of the very highest grade. Its prize record is unique in both the breeding and fat stock sections. In 1896 Minx of Glamis was a double champion at Birmingham and Smithfield. In 1898 Ju Ju repeated the same feat. In 1899 and 1900 the herd won the Queen's Challenge Cup outright at Smithfield for the best animal bred by exhibitor. In 1901 there was another repeat at Birmingham and Smithfield—Brunhilde being the name of the champion; and again in 1902 Layia did it over again. The success of Glamis at these shows is all the more remarkable—coming as the cattle do from Angusshire which had never been much in the fat stock show limelight. It simply shows, that wherever the man is, the cattle will be just as ready to respond to that man's master force and touch.

The Kinochtry and Pictstonhill herds of W. S. Ferguson, son of the late Thomas Ferguson, are impressive examples of persistence in which much reward has been found.

The Portlethen seems still to rank as the oldest herd in existence. The very names of the animals one reads in the pedigrees have another century appearance and sound. In many ways the herd epitomizes the history of the breed. It traces beyond, through the herd of Mr. Williamson of Portlethen Mains—another reminiscence of the name of St. John's Wells—into the same century in which men of the same name were creating the solid foundation for the breed. The herd first appeared at the Highland Society's Show at Aberdeen in 1834, when it won second for cows to another Mr. Walker—of the Fintray family. At the Centenary Show at Edinburgh, 1884, the herd then in possession of its present owner, George J. Walker, was awarded a gold medal for a set of triplets, Asia, Africa and America, daughters of Alexandrina (894). J. G. Walker states that the first bulls he can trace as used in the herd came from Wester Fintray and Fernflatt—associated with Panmure fame—that is, 1826 to 1836. The first "crack bull" was perhaps Banks of Dee, which was followed by the Andrews and Raglan for which the late Emperor Louis Napoleon offered $1,150 at the Paris Show in 1856; Fox Maule, a rare animal, indeed, and Palmertson, a straight descendant of the Wester Fintray herd. The first regular Herd Book of the stock was dated 1840, and one of the first animals sold to go out of the country was a cow, Duchess, that went to America—in 1850.

Like the Walkers of Wester Fintray, the Portlethen Walkers paid great attention to the milking qualities of the herd—keeping accurate records, as referred to in the chapter on Milking Qualities. A study of the old sorts in Portlethen is as interesting as anything the breeder could indulge in—for the reason of the existence of so many descendants of the herd. To show how the best things going in the breed were noted by the breeders keeping their eyes open for the same, take Matilda Fox—dam of the dandy Fox Maule. She was bred by Bowie, Mains of Kelly, and was picked up by Mr. McCombie of Tillyfour; from whom she was bought in 1857 at the Bridgend sale by the late Mr. Walker, to produce "one of the most renowned bulls of the breed," Fox Maule, winner of the senior honors at Dumfries in 1860. This interesting account of Portlethen may be rounded out by reminding the reader that Luxury, the Smithfield champion, originated in a family long noted in the herd.

The Spott herd of William Whyte is also one of the oldest in existence and noted for its Jipseys, a member of which Ju Ju, exhibited by the Earl of Strathmore, Glamis, was champion at Smithfield, 1898. Jipsey Baron, breed champion at Inverness, 1901, was also of the same family.

The herd of Col. Chas. McInroy, of The Burn, dispersed in 1920, is another of the herds running into the prehistoric period of the breed—like Portlethen and Aldbar—and its quality is proved by the fact that from it came the Smithfield champions of 1905 and 1909—Burn Bellona 35998 and Pan of the Burn 27244.

The Wester Fowlis herd of Alexander Strachan promises well to assume a traditionary standing in the great Vale of Alford as the hereditary herd. Wester Fowlis and Bridgend are in the same parish, Leochell-Cushnie, and the late Mr. Strachan and Mr. McCombie had many a deal together. The Wester Fowlis—and other local herds—benefited by the overflow from Tillyfour. Mr. Strachan, the present owner, seems to have been endowed with the instinct of his father—shared in by the younger, Pat Strachan, who is also breeding. The herd produced Coronal, dam of the Scottish champion (Edinburgh Fat Stock), 1916, which was reserve for the championship at Smithfield the same year.

The Ardhuncart herd of the late William Walker appears to have been continued by the son—it dates back to the prehistoric period likewise; while that other herd, that of James Walker, Westside of Brux, has had the distinguished honor of providing material for the Royal herd at Abergeldie.

While in the old Aberdeen-Angus districts of Scotland, purebred herds have usually been kept up even when the estates have passed from one family to another and the old herds that made history from a quarter to a half century ago were dispersed, this is not always the case. The Ballindalloch herd of Sir George Macpherson Grant, the third generation of the line since Aberdeen-Angus breeding operations were laid on a modern scale to usher in the "Ballindalloch influence" to change the breed the world over, is still the leader. A close second must be rated the Harviestoun herd of Mr. J. E. Kerr, whose scientific and practical breeding work evolved the "Juana Ericas" which won an unprecedented string of Highland championships with its females, and whose bulls have been tops and champions at Perth shows and sales so regularly in recent years. Bleaton, founded by Messrs. Marshall and Mitchell but a decade and a half ago, headed by Baron Beauford, sire of Etrurian of Bleaton, double champion at the Royal and Highland shows for two successive years, takes rank close behind these two. Doonholm, breeder and developer of so many high-class females in recent years; Skillymarno, where the late Mr. Chas. Penny gathered a carefully selected and high-priced herd that was turning out bull and female prize winners, Inchgower, Mounthooly, Ballintomb, The Banks, Dundas Castle, Connage, Hatton Castle, Glenfarcles, The Dell, Mulben, Tillyrie, Auchterarder, Hayston, Dupplin Castle, Advie Mains and Carr-Bridge (both owned by Peter Grant), Aberlour, Dalmeny, Morlich, Castlecraig, Kinermony, Dandaleith, Candacraig, Philorth and Careston Castle. In England, Maisemore, Stagenhoe, Claverdon Leys, Langshott, Bywell, Morden House, Goodwood, Monks Horton, Conholt Park and Buckland herds appear to be the present day leaders. In Ireland, the Lisnabreeny, Curragh Grange, Lisard, Gortnaskehy and Tubberdaly seem to be the present day leaders.

TRIUMPHS IN FOREIGN COUNTRIES

About the date of the advent of Pride of Aberdeen, and having well established itself as a national breed, it was felt that a trip abroad would extend its fame, and its first foreign exhibition was at the Paris International Exhibition of 1856, where it was shown in considerable strength. This was the show at which Charlotte and Hanton, shown by Mr. McCombie, made such an impression on the judges that they wrote the following:

"The specimens of this breed possess the following characteristic points: Perfect homogeneity of race, beauty, richness and regularity of form, softness of skin, mellowness in handling; the whole united to a muscular system sufficiently developed. They presented, besides, a considerable mass of flesh, supported by a comparatively small volume of bone. We are aware, besides, that the breed joins sobriety to a great aptitude to fatten and that it supplies the butcher's stall with beef of much esteemed quality; that it produces milk in satisfactory quantity, is of a sweet temper, and is also endowed with prolific qualities."

Dutrone, writing twenty-two years later, after the crowning event of 1878, also referred to this exhibition of polls: "I doubt whether those he brought out at the last International Exhibition in 1878 were of equal merit. I well remember the laudatory and wondering remarks of foreign visitors when passing around the stalls where the stately masses of the polled cattle were drawn up in a black and imposing array, even and level, as if the chisel of the sculptor had been plied over their grandly fleshed frames." The last part of the foregoing is what we, of course, have pleasure in noting—being that which characterizes the breed upon all "dress occasions."

In the grand march past of the breed the International Fat Stock Show of Poissy, 1857, is likely to be overlooked. Dutrone records the fact that "out of six prizes offered for polled oxen, Mr. William McCombie of Tillyfour obtained four." One of the Tillyfour steers proved to be the heaviest of all breeds, when weighed by order of the Emperor. He was a son of Bloomer, Mr. McCombie's favorite cow.

The International Exposition at Poissy, 1862, created a worldwide sensation, only equaled by that which occurred in 1867 (Black Prince's year) and in 1878. At this Poissy Show Mr. McCombie won the highest honors with the steer that came to be known as the "Poissy Ox" or the "Mammoth Ox." His girth was 9 feet, 8 inches—having "come out" four inches since his last appearance at Birmingham and Smithfield, four months previously, where he had gained the first prizes; his length from tail to chime was 5 feet, 6 inches, and from tail to poll 7 feet, 9 inches; his height was 5 feet, 1 inch, and length below knee 9 inches "round." He weighed 1,250 kilos (2,750 lbs.). He was sold to the Emperor's butcher and a photograph taken of him for preservation. Mr. McCombie was also first in the younger class of steers, and first in both classes for females.

In 1867, the great Steer, Black Prince, from Tillyfour, carried all before him at the fat stock shows, crowning his career with the grand championship at Smithfield. That has been honor enough for most breeders, but Mr. McCombie was more fortunate. Black Prince, so greatly impressed the late Queen Victoria that she "commanded" his attendance at Windsor for her personal inspection. Later, after the black polled champion—the first that ever won the purple rosette at Smithfield—had joined the great majority, Queen Victoria visited Tillyfour, an imperial honor conferred on but mighty few of her loyal subjects in any walk of life.

Eleven years later the fame of the Aberdeen-Angus breed was destined to be blazoned across the whole wide world. To the Universal Exposition at Paris, sixteen doddies were dispatched from Scotland, McCombie contributing nine, Sir George Macpherson Grant of Ballindalloch, six, and George Bruce, of Tochineal, one. The victory of this little band of blacks was thorough and complete. They swept the boards of all the prizes for which they were eligible to compete, winning among other honors, the grand championship for groups of foreign-bred cattle and the supreme championship as the best beef producing animals on the grounds. This final struggle was with the white Charolaise breed of France, the bovine display in black and white having been described as the most picturesque ever witnessed. This sweeping triumph set the seal of fame upon the Aberdeen-Angus breed, its conquest of the world's markets dating from that memorable season.

IN ENGLAND AND IRELAND

In 1875, a herd of Aberdeen-Angus cattle was founded in Sussex, in the south of England, and the year following the Langshott and the Duke of Grafton's herds were given being. In 1880, Dr. Clement Stephenson, who was to carry the black and all-black banner far to the front, started the Balliol College herd in the northern borderland. Its great success with the Smithfield champions, Luxury in 1885, Young Bellona in 1887 and Benton Bride in 1894, made this herd's fame national and when Abbess of Turlington, daughter of the Abbess of Benton, won the supreme championship at the Columbian in Chicago, 1893, international renown was accorded it without stint.

Invasion of the Emerald Isle first took place as early as 1864, but owing to the disturbed condition of that country, little progress was made for years. However, within the past two or three decades great strides forward have been made, there being now about as many breeders of Aberdeen-Angus cattle in Ireland as there are in England.

The growing importance of the breed in Great Britain is proved by the establishment of the English Aberdeen-Angus Cattle Association in 1899—the first honorary secretary being Mr. Albert Pulling, Beddington, Croydon, who has published several books relating to the breed. Again in Ireland, the Irish Aberdeen-Angus Association was formed—the honorary secretary of which was Mr. Wickham H. B. Moorhead, Carmeen, Newry.

THE SCOTCH HERD BOOK

It was after the Aberdeen Highland Society Show in 1840, with the rising wave of interest and enthusiasm which then appeared, that the idea of a herd book was first mooted. Collection of records and data was begun in 1842, deposit being made in the museum of the Highland and Agricultural Society of Edinburgh. A severe setback was encountered in 1851 when the whole material brought together was destroyed by fire. A fresh start was made in 1857 at the request of some of the leading breeders, and Vol. I appeared in 1862. Vol. II was produced in 1872, and in subsequent years Vols. III, IV, and V, all by private enterprise. Then, after the Paris Exposition of 1878, upon the initiative of Sir George Macpherson Grant, the Polled Cattle Society was formed, the first general meeting being held in 1880.

Vol. VI was issued by the Society, and in it only nine breeders south of the border had entries. At present there are about 600 members of the Society. Galloways were included in the first four volumes of the Herd Book.

The Polled Herd Book was the name originally chosen, which name remained after the exclusion of the Galloways, for the registry of cattle of the "Abedreen and Angus breed," but in 1886, following the example of the American Association, the breed became known to the whole world as the Aberdeen-Angus and the British book was rechristened accordingly.

SMITHFIELD SHOW AT LONDON, ENGLAND

Dr. Clement Stephenson of Balliol College Farm, Northumberland, was the first great modern winner with Aberdeen-Angus cattle at Smithfield. But there were victories between that date back to 1867; and there had been Scotch cattle exhibited at Smithfield as early as 1805. The first mention of Scotch cattle in the records of this great show was in that year and Scotch cattle were exhibited in the following years: 1807-1811, 1825-1828, 1830-1832, 1836, 1840-1843, 1846-1851, 1852. From the last named year onward there were more or less regular exhibitions of Scotch cattle; but it was not until 1856—the year of the first Paris breeding show—that there was a separate class for Scotch polls—which included both Scotch breeds.

The total number of prizes awarded up to 1851 for Scotch cattle was 43, for which there was awarded $2,500; all prizes being for steers or oxen. From 1862 to and including 1867 (Black Prince's year), the Scotch won two silver cups—in 1865, a Scotch horned ox, exhibited by the Duke of Sutherland, and in 1867, the Tillyfour poll. In 1866, the silver cup for the best steer or ox was won by a Shorthorn-Scotch Polled (Aberdeen). It has been recorded that one of the Williamsons sold an Aberdeen ox which was exhibited at one of the shows previous to 1828 at least. Then in 1830, Mr. Watson sent his famous heifer and also later another steer. It was not until 1859 that Mr. McCombie exhibited at Smithfield, when he gained in that and the succeeding three years, and in 1864, 1867, 1868, 1870, 1871 and 1873, 1874 and 1875, the cup for the best Scot—besides, of course, the greater victories in 1867 with Black Prince.

In 1871, James Bruce, Burnside, Fochabers, won the female championship at Smithfield with an Aberdeen-Angus three-year-old heifer, bred at Mulben. The following year the same exhibitor sent forward the champion in a three-year-old steer bred by John Macpherson, who (on Mr. Paterson's death) had become tenant of the now historic Mulben. In this same year Mr. McCombie had won the championship at Birmingham. These two champions, of a comparatively youthful age, as ages were then considered, were thus the first specimens of the breed to shatter forever the old-time prejudice—that a polled Scot, or Aberdeen-Angus rather, could not mature early.

Aberdeen-Angus cattle were "unwelcome intruders" in England and for awhile they only had an occasional nibble at the big things at the fat stock shows. Then the crowning victory of this, the second Golden Age, occurred in 1861, when Sir William Gordon Cumming, of Altyre, with a pair of polls little more than two years and a half old, carried off both the male and female championships—the heifer at last being awarded the highest honors and the steer standing reserve to her. Again in 1887, Altyre exhibited the champion steer. In 1901, 1903, 1916, and 1920 the best steers were shown by J. J. Cridlan, Maisemore Park, Gloucester.

Females have rendered great service to the breed at the fat stock shows. Just recalling the champion of 1871 and 1881, we find the female championships won as follows: 1884, by John Strachan, Montcoffer, Banff; 1885, Clement Stephenson (also grand championship); 1887, by the same, and again in 1889 and 1894 the same—the winner in the latter year being grand champion both at Birmingham and London. In 1893, the winner was J. Douglas Fletcher, Rossshire. Then in 1896, 1898, 1901 and 1902, Lord Strathmore, Glamis, won the female championship, each one of these being also the grand champion. In 1897, Lord Rosebery won with a heifer bred in the Vale of Alford, and again in 1906 and 1907 repeated the trick with Dalmeny home-breds. In 1905, Col. McInroy won with Burn Bellona, which was also grand champion, and in 1913, the Duke of Portland scored supreme victory with Beauty of Welbeck. In 1910, 1914 and 1916, J. J. Cridlan's Maisémore herd furnished the grand champions. For 1919 a cross-bred three-fourths Aberdeen-Angus yearling heifer won for Major J. F. Cumming.

Since the end of war has brought the Smithfield show back to its pre-war standards as to quantity and quality of entries, there has been a swing to black and blue-gray winners that has fairly swept opposition off its feet. At both the 1920 and 1921 shows, Aberdeen-Angus won both championship and reserve, the three-quarter blood Aberdeen-Angus that was champion at the 1919 show being reserve to a pure-bred in 1920. At the 1921 show Aberdeen-Angus pure-breds won both the highest and reserve honors, and this victory was practically a duplicate of all the other British shows of note in both Scotland and England—Birmingham, York, Norwich, Edinburgh and Aberdeen. The Aberdeen-Angus has broken down the old barriers of prejudice in the strongest territory of its rival breeds and their followers and propagandists.

ABOUT THE CROSSES

In the Prime Scots, of course, we include Aberdeen-Angus crosses, either way. Speaking of these, Mr. Robert Bruce, well known in the leading live stock circles of both countries, said: "The Aberdeen-Angus-Shorthorn cross is highly valued by northern breeders, and the larger numbers of farmers in England and Ireland, who have resorted to this cross prove pretty conclusively the general appreciation of the many good qualifications belonging to the blend. Where ordinary judgment is exercised in the selection of sires and dams, the excellence of the produce is at once assured, as the blending of the Shorthorn and Aberdeen-Angus blood results in a quick feeding and rent-paying one. A glance at the records of the great fat stock shows at once indicates the important position taken by these Shorthorn-Aberdeen-Angus crosses in the annual prize award lists. There has been a widespread demand for Aberdeen-Angus bulls for crossing purposes all over the north of Scotland, and this system of crossing has also made its way into other portions of the kingdom. In my opinion it is immaterial how the cross is brought in, whether through the Shorthorn sire and the polled cow, or the polled bull and the Shorthorn cow. Circumstances and situation may alone be left to guide the breeder in the selection of the sire to use."

THE BREED'S RISE IN AMERICA

Though perhaps the first Aberdeen-Angus animal that ever trod American soil was the cow Duchess which went from Portlethen in 1850, it was not until 1873 that stock was imported for the express purpose of improving the range cattle. In that year the late George Grant of Victoria, Kansas, imported three bulls, two of which he exhibited at the Kansas City Fair—the first polls that ever appeared in an American showyard. These bulls, which created much interest, were the forerunners of the great influx which occurred a few years later as a result of the world-wide renown the breed had acquired by winning the two champion group prizes in Paris in 1878.

These bulls were used upon the common stock of the range, horned and coarse, and they changed the "complexion" and appearance of the old stock. Many half-breed steers from these Aberdeen-Angus bulls were fed by a number of feeders and gave them a foretaste of the quality that lay beneath the black skins. In 1876 James MacDonald, the late secretary of the Highland and Agricultural Society of Scotland, visited the ranch and reported them as doing splendidly. In 1876 there were probably more than 800 black polled calves after them, declared to have been superior to any ever seen in those parts before. They were short-legged, big around the girth; vigorous, healthy and thoroughly at home; they proved themselves superior in every way to the Shorthorn and other crosses; standing the winters better, coming out in remarkable condition, without the necessity of artificial food or coddling as the other breeds required. It is a pity Mr. Grant did not live to reap the benefit of his foresight, which would have been his in good measure. Yet his work followed after him. In 1883, there were sold in the Stock Yards at Kansas City, fourteen half-breed Aberdeen-Angus steers, the produce of the Grant bulls. They were bought by Charles Still, at $4.25, averaging 1038 lbs. in weight. Four months and six days later, they were sold at the same yards for $5.45, when they averaged 1280 lbs. in weight, and then they were not "full-fed."

The inquiry from America had just begun after the Paris Exposition. Perhaps the first to seriously inquire into the opportunities for importing the breed into the United States, was Mr. Libbey, then editor of the "Rural New Yorker," who visited Scotland one summer and made an investigation of the breed. Then John Wallace, publisher of the American Trotting Register and "Wallace's Monthly," wrote to Tillyfour about them, having become interested with his friend, Mr. Redfield, Batavia, New York—one of the first importers. But it was not until after the dispersion sale of the Tillyfour herd that the rush—the boom—began.

What might be termed the parent herd of America was that formed by Anderson & Findlay, Lake Forest, Ill. Mr. Findlay was indeed a native of Buchan and had retained all the affection for the native "humlies" that everyone acquainted with them in youth undoubtedly is bound to preserve. It was during the summer of the memorable year 1878, that Anderson & Findlay commissioned Mr. Findlay of Peterhead, Scotland, brother of the latter, to purchase five heifers and a bull from the best herds; which commission was followed by others. Anderson & Findlay exhibited their importations at the Illinois and other fairs.

Among the animals in their first importation was a bull from the old Westside of Brux herd, descending from Keillor blood. In the next importation visits were made to Mains of Kelly, Wellhouse, Bridgend, Earnside and other herds; resulting in the securing of specimens of the Jennets—derived from old Young Jenny Lind, a Tillyfour foundress; Lady Jean, a good old Rothiemay sort; Montbletton Charlotte, Westertown Victoria and other families. Waterside King 2d of the old Fanny of Kinnaird tribe, and Basuto, a Blackbird-Erica bull, headed the new herd. Again, in 1882, Scotland was revisited, and the herd of Burleigh & Bodwill, that had been formed at Vassalboro, Me., was purchased, in which were Ericas and specimens at Vassalboro, Mains of Advie, in other tribes. The verdict of the owners of the herd which was the foremost in the country for years was that "for plains and beef cattle, early maturity, weight, quality of beef and hardiness they cannot be surpassed."

In 1881 the herd was conspicuous at St. Louis, in 1882 had the championship for the best cow at Kansas City, and it also spread the fame of the breed at many other fairs. From this fountain-head many American herds were supplied—notably those of the late T. W. Harvey, Turlington, Neb., and J. V. Farwell, Chicago, who was interested in land development in the Panhandle of Texas. The influence of the Lake Forest herd was spread wide athwart the country. On the X. l. T. Ranch, which was the scene of the great drama played by the Lake Forest sires, experience fully demonstrated the value of the breed as range transformers—just as the Victoria bulls had done in Kansas. On the other hand, the Turlington herd, going into the fight of the breeds at the fairs and fat stock shows, did more perhaps than any other to break down the barriers and make easier the path of the feeder and exhibitor who came after Mr. Harvey's death.

In the same year F. B. Redfield, Batavia, New York, established his herd; his purchases were made at Kinochtry, being three heifers and a bull—all by Shah—senior male champion at Dumfries, 1878. The females were of families deep in Keillor blood. Two years later sixteen animals from the same herd were imported—nine bulls and seven females, mostly of the same blood. This herd made an enviable reputation in the great showyards and the bulls carried the stamp of the market-topper to the ranches where they were introduced. Mr. Redfield's estimation of them is summed up in a word—"they have constitutions of iron."

In 1881, J. J. Rodgers, Abingdon, Ill., founded a herd by selections from Kinochtry—of Favorites, Baronesses, Princesses and other families, six of them being by Prince of the Realm, Shah's son.

In the same year, Messrs. Gudgell & Simpson formed their herd, Col. Simpson visiting Scotland and personally selecting the foundation stock. His selections were made from Waterside, where he got Blackcap (4042) bred at Ballindalloch—an Erica-Mayflower, Rosa Bonheur 2d (3531) bred at Tillyfour, and others of Drumin, Greystone, Old Morlich, Mains of Advie and other noted tribes. The bull selected to head the herd was Knight of St. Patrick, from Bridgend, which had a very creditable career in this country, siring some of the most noted animals that appeared at the early shows. In 1887 this herd was sold to the Fairmount Cattle Company, Stratton, Neb., and it also made an excellent record in the ring.

In 1882, A. B. Matthews, who had already secured some animals from Canada and elsewhere, visited the home of the breed and made an excellent selection from Haughton, Greystone, Waterside, Kinochtry, Easter Tulloch, Gavenwood, Baads, Balquhain and Blairshinnoch. His herd, when he had it all assembled, numbered 170 head. Mr. Matthews was a prominent figure in breed circles and at the showyards and sales rings for a number of years succeeding his entry into the ranks of the importers. He wrote early in his experience: "The prospect for the breed is beyond anything that I have ever known for any class of cattle."

Another great exponent of the breed of this era was George W. Henry, also of Kansas City, whose visit to Scotland is still remembered with pleasure. He visited Bridgend and secured Dandy 2d (3266), Empress (Queen Mother family) and other old-fashioned sorts. Greystone, the old herd of James Reid (now rebuilt and owned by Col. Harry Forbes), so well known to all breeders, supplied Bella 2d which was later sold by Mr. Henry for $1,000. Other selections were also made from Wester Fowlis, Blairshinnoch, Wellhouse and elsewhere. The bulls included Black Commodore from Montbletton—a Ballindalloch Coquette. Mr. Henry's appearance in the fat stock show arena is mentioned in a succeeding chapter.

This year, 1882, also saw the founding of the herd of Estill & Elliott, Estill, Mo., which had a marked effect in forwarding the fortunes of the breed. Among their purchases were Effie of Aberlour, at the price of $2,400; Carrie of Montbletton, May of Achlochrach and Harriet of Balfluig, from which were bred the state fair winners of the herd, which was dispersed in 1900, when 58 females averaged $583, and 14 bulls $561. Lucia Estill brought $2,800 from the late W. A. McHenry, of Iowa, then a new star on the horizon. Purchasers from eleven states took home animals from this noted herd that had stood in the front of the battle line for the breed.

A special place in the history of the breed will always be reserved for George Geary, who, with his brother, started his career as a breeder and importer in 1882, purchasing nineteen head from Gavenwood of the strongest families maintained there. Representatives of Ballindalloch, Balliol College Farm, Kinochtry, Easter Tulloch, Westertown, Montbletton, Rothiemay, Drumin and Queen Mother families were chosen. In 1886 they startled the breeders by the purchase of the entire Gavenwood and Rothiemay herds numbering fifty-eight and thirty-four heads respectively. The story of Geary's immortal Black Prince of 1883 is given in the next chapter.

The great year 1883, also saw the advent of the Heatherton herd of the late Goodwins, John S. and W. R. This year (1883) was perhaps the banner year as far as importations went, for it was estimated that in it 800 animals were imported from the old country to join those in America and form new centers of the breed. The late Campbell Macpherson Grant, brother of Sir George, sent over altogether 250 head, commissions for various breeders, and that was perhaps the largest number ever sent over by one man in one season.

In this year, Leonard Brothers, Mount Leonard, Mo., founded their herd—from purchases made for them by that commissioner. Two years later Mr. Leonard was exemplifying the merits of the breed as the market-toppers, having sent a lot of sixteen steers to Chicago, weighing an average of 1,593 pounds, and selling for a record price.

John D. Larkin of Buffalo, New York, with farms at Queenston, Ontario, Canada, in 1910 imported nearly 200 head, which is probably the record importation of Aberdeen-Angus in any one season by one breeder. The Eschers, Stoner & Baird, of Iowa, and Carpenter & Ross, of Ohio, also made large importations in recent years.

As time goes on the Turlington herd stands out more and more prominently in the perspective of the past. T. W. Harvey, its owner, lavished his resources on the breed—not in a wasteful, but in a thoroughly constructive manner. He determined to have the best—that the Aberdeen-Angus should come into its own. And but for him the hey-day of the breed might have been longer in dawning than it was. He was fortunate in enlisting the services of William Watson, son of the late Hugh Watson, of Keillor—and if there was ever anyone who would have gone through fire for the breed it was "Uncle Willie," as he was affectionately called.

The Turlington herd was sumptuous—superb! It contained thirteen Heatherblooms, six of the Bride family, seven Jeans of Easter Tulloch, three Nightingales from Waterside, one Kinnaird Fanny, four Alexandras of Montbletton, three Queen Mothers, two Matildas of Waterside, three of the famous Abbesses, six Beauties of Glamis, five Victorias of Westertown and Balwyllo, four Waterside Minnies, three Keillor Favorites and Princesses, two of the Progress family, three Easter Tulloch Margarets, six of the Carnation tribe of Corskie and South Ythsie, two Hecubas, two Duchesses of Shempstone, one Evelyn of Fintray and four of other sorts. The sires at the head of the herd were Guido, a Kinochtry Favorite by Young Captain (4238), dam Beauty of Brucehill (742); Black Knight, bred by Gudgell & Simpson, sire Knight of St. Patrick, dam Blackcap (1442); Eurymedon, an Erica by a famous Balliol College Farm Souter Johnny; and Errant Knight, another Erica, by Sea King (1450), dam Errantine (4642). The breeding herd carried off the grand championship at Nebraska State Fair in 1884, when headed by Guido and Waterside Minnie.

Thus Turlington was the home of the Heatherblooms and of Abbess of Turlington, with which McHenry carried away the champion honors over all beef breeds at the Columbian Exposition in 1893. A purchase by Mr. Harvey from Bradley Hall, Antelope, became also the dam of the first International grand champion steer, Advance, exhibited by Stanley R. Pierce in 1900. The victories of the herd in the fat stock shows are recounted in the section dealing with "The Breed in the Arena."

In 1883 also, J. J. Hill, the railroad magnate who tried to benefit the farmers in the Northwest by spreading good Aberdeen-Angus bulls in that territory, obtained from Clement Stephenson, Balliol College Farm, seven young bulls of such quality as Plum Pudding and Patrician of the Mulben branch of the Prides of Aberdeen; the twins, First Flight and First Foot, of the Boghead Flora family; Busar of the Rothiemay Heather Bells; Spice Box of the old Bognie Southesks, and Advocate of the famous Abbess family, specimens of the latter of which had been previously sent to Turlington. Mr. Hill was an extensive importer and successful exhibitor at the great fat stock shows for many years.

The herd of J. H. Rea & Sons, Carrollton, Mo., founded over 40 years ago and now owned by J. W. Rea, is one that has carried on in an unbroken line of breeding, feeding and importing since its foundation. It inaugurated a new era in American cattle feeding when it sent the first bunch of fat steers to market carrying Aberdeen-Angus blood. This group of cattle, marketed at Chicago and then sent east to New York City, astounded the stock yard and slaughter interests and fixed the spotlight on the breed's produce for fattening for the best markets.

The Woodlawn herd of the Pierces, the late B. R. and his son Stanley, now owner of this unbroken line of breeding, has had a marked effect on the breed's direction in this country. Advance, the first single steer grand champion at the International Live Stock Exposition, gave the breed a great send-off with the inauguration of the Chicago International and that record has been faithfully followed up by breeders and feeders. The public and private sales of the Woodlawn herd have scattered many cattle over the Central West and even Canada and South America.

The Highland herd of the Campbells in Minnesota is another that has done much sound constructive work in scattering the breed throughout the Northwest in the past 40 years. Its importations of the Morlich bulls, Juba and Eclipse 2d, and its early showing and feeding gave it a strong place in the Northwest. It also produced carcass and single grand champions for the Chicago International.

BECOME PERPETUAL CHAMPIONS IN U. S. ARENAS

The late George Geary was the first to realize that the breeders would have to do something in the fat stock show ring. He was able to secure the steer Black Prince, bred in Aberdeenshire, which had stood second at Smithfield in 1882, when shown by Mr. Lowthian Bell, Yorkshire. He was shipped from Liverpool, the next year, when he weighed over 2,500 lbs. When he reached Kansas City after a fearfully rushed journey—which had to be made by "express" at a cost of $400 to be in time for the show, he dropped to 2,360 lbs. After such a fearful strain he was not shown in very fit trim, and did not realize the hopes of Mr. Geary and was comparatively overlooked. But at Chicago he gave the breeders a foretaste of the quality of the breed by winning the championship prize given by the butchers—always the friends of the Aberdeen-Angus as the best three-year-old, beating in this notable contest the steer that had been placed over him at Kansas City, as well as the steer that was later awarded the open grand championship at the same show. Next year, 1884, he appeared again; and when slaughtered his carcass dressed 71.3 per cent net to gross.

G. W. Henry, Kansas City, one of the pioneer importers, as we have seen, was also a demonstrator of the breed at the Kansas City Fat Stock Show. In 1884, he exhibited an animal that made a great impression on the breeders—Bride 3d of Blairshinnoch, which easily won the prize as the best cow of any breed, and when slaughtered won the first prize in the dressed carcass competition. Alive, she weighed 1,395 pounds; her carcass, 881 pounds, being 65.15 per cent of dead to live weight. She got no pampering, indeed being taken from the pasture, where she received all the feeding she got—and that was little. At the same show, an Aberdeen-Angus-Hereford grade, winner of the gold medal for the best beef animal bred by the exhibitor, weighed 1,615 pounds, being 665 days old, showing the satisfactory daily gain of 2.43 pounds.

At the same show, Kansas City, 1884, the Indiana Blooded Stock Co. won gold medal for the best beef animal bred and fed by the exhibitor, with the heifer Burleigh's Pride, an Aberdeen-Angus-Hereford cross, whose weight for 665 days was 1,613 pounds, a daily gain of 2.43. The same exhibitor's yearlings were first and second for early maturity in a class of ten of the different breeds. The American Aberdeen-Angus Breeders' Association also showed a two-year-old, which, second in its class on foot, was first in its class for early maturity.

Benholm, exhibited by J. J. Hill, the railroad magnate of the Northwest, at the great 1885 Chicago Show, is another landmark. He was then two years old and dressed out 71.4, which rather opened the eyes of the breeders.

Again at Kansas City, 1886, appeared Sandy by Knight of St. Patrick, "which had a wonderful career," shown by Messrs. Gudgell & Simpson. He weighed 1,470 pounds at 393 days old, showing a gain of 2.47 pounds per day. He won first prize as a yearling at the American Fat Stock Show at Chicago, 1885; next year, the Breeders' Gazette's gold shield for the best in the show bred and fed by the exhibitor, and the championship of the entire show over all breeds, grades and ages. At that time he showed an average daily gain of 1.97 pounds. In 1887, at Chicago, when he headed the sensational Turlington herd, won the grand championship, he weighed 2,225 pounds at 1,322 days old, which gave a rate of 1.68 pounds gain per day.

In 1887, we come to Turlington's year, when T. W. Harvey, that great king of doddie men, made such a fight for the breed that has never been forgotten. He, aided by "Uncle Willie" Watson, brought the doddies forth in grand style, bearing the black standard aloft in the thickest of the fray and carrying terror into the ranks of the breeders in the contending camps. Then it was that "The Black Watch" was on guard—and it was well they were. But, notwithstanding all their vigilance, this year went down into history as the "should have been Aberdeen-Angus year." Yet one glorious victory was won—that of the grand champion herd.

Mr. Harvey had been victorious all along the line at Kansas City, winning the championship there over all breeds with Black Prince of Turlington, and also the herd championship and many other prizes with Black Prince, Tam O'Shanter and Alexander Knight.

Coming to Chicago there was a harder task laid out for his "intruders," but though the fortunes of war went against him in the single steer championship, he nevertheless swept the field when it came to the herd championship, which proved one of the most sensational witnessed on American soil. His crack herd consisted of Sandy, son of Knight of St. Patrick, Black Prince by Guido, Tam O'Shanter, son of Black Knight (a son of "St. Pat") and Alexander Knight, by Black Knight also. It was a phenomenal exhibit, but the judges split and John C. Imboden, Decatur, Ill., was called in as referee. He ordered the contesting herds to draw up in a row, red and black alternately, and placed the blacks first. It was a daring award at the time, but after it was made the doddies were established in the front rank.

Mr. Harvey's winnings at Kansas City had been $2,045; at Chicago $6,185—which figures show the magnitude of his victory. And the doddie men were listening also to the news from across the sea—from Birmingham and Smithfield—Young Bellona was the double champion, and to her at Smithfield another Aberdeen-Angus stood reserve. In this memorable year of 1887, also came out the wonderful white-legged steer Dot, shown by Wallace Estill, who sold him to Mr. Imboden, and that Illinois feeder won the grand championship with him the following year. Dot weighed 1,515 pounds when 863 days old—equal to a daily gain of 1.75 pounds. He dressed 1,040 pounds—equal to 69 per cent of his live weight.

In the practical exhibitions, dressed to live weight, percentage of daily gain and carcass tests, the Aberdeen-Angus also this year proved themselves true to their title as prime Scots and the premier beef breed.

No history of the breed, however, condensed and brief, would begin to tell its modern story without a generous paragraph or so on the work of the colleges and experiment stations of the United States. This phase of breeding and feeding development came into existence with the launching of the International Live Stock Exposition and the part that the colleges and experiment stations immediately began playing in the Chicago show, Michigan, Iowa, Minnesota, Nebraska, Kansas, Indiana, Missouri, Ohio, Pennsylvania, California, (during the last few years since the beef herd was founded), and to a lesser extent Wisconsin and North Dakota. The most consistent work has been done by Iowa, though Nebraska, Minnesota and Michigan stations have accomplished wonders. Iowa, beginning in 1902, has made annual showing, winning four single steer and six beef carcass grand championships with its Aberdeen-Angus steers, as well as several of the groups, such as steer herd of three and "Best 5". With the larger of these agricultural institutions, it has been possible to put in foundation herds of the leading breeds of live stock—beef and dairy cattle, swine, sheep, horses and poultry. Now every agricultural college either has started or plans to follow the example of these larger institutions that have blazed the way in breeding and feeding the best their resources will afford. In such an atmosphere of unprejudiced science with the ambitious youth of the land to absorb the lessons of breed adaptability, the Aberdeen-Angus has been wonderfully stimulated.

59 GRAND CHAMPIONS AT CHICAGO

But it was the International Live Stock Exposition that "put the Doddie over" with the cattle feeders of the United States. Founded in 1900, at Chicago, by a little group of far-sighted men, it brought together all the great interests that center about our live stock business—rancher, feeder, breeder, packer, commission man, and for good measure and idealism, college and experiment station.

The International has now been held for twenty-two years, 20 shows in that time having passed, foot-and-mouth disease causing the abandonment of the shows in 1914 and 1915. In this time, the Aberdeen-Angus have won 59 inter-breed steer grand championships in the four principal classes—Single Steer, Steer Herd, Carlot and Carcass—while the Herefords were taking 8, the Shorthorns 6, crossbreds 4, and a mixed herd containing one Aberdeen-Angus, one Shorthorn and one Galloway, the other one grand championship.

In single steers, Aberdeen-Angus have won outright 12 grand championships and a 13th was a cross-bred Hereford-Aberdeen-Angus. In steer herds of three, there was no class the first two years, but at the eighteen shows since, Aberdeen-Angus have won 12 grand championships to two for the Shorthorns, one for Herefords and one for the mixed trio that contained one Aberdeen-Angus, and one trio containing 2 Shorthorn-Aberdeen-Angus and 1 Hereford-Aberdeen-Angus crossbreds. In the classic fat carlot show, Aberdeen-Angus have won 16 out of 20 grand championships, and one of the other four was given the Herefords after the Aberdeen-Angus had been disqualified on an age technicality. The Shorthorns won one carlot grand championship.

But it has been the carcass shows that have given the Aberdeen-Angus backers their question-slogan, "What's under the hide?" For 19 successive shows, the Aberdeen-Angus steers have won this grand championship over all breeds, grades and crosses, a grade Shorthorn winning it at the first show in 1900.

That first International was a rather crude affair when the present stupendous and smooth-working exposition is seen. The Breeders' Gazette, commenting on that first carcass show, states that the judge did not stick to the usual carcass standards and paid no attention to marbling. At any rate, the International has been entirely "black," as far as grand championships are concerned, ever since. As this is the most practical class of all, one butcher expert judging the animals on the hoof and a second passing on their steaks and roasts as they hang in the cooler after slaughter, the question of the supremacy of the Aberdeen-Angus as a butcher's beast is no longer even argued. The marbling of fat and lean throughout the carcass is an inherited quality that no other breed has successfully imitated even closely, while dressing percentage, fine quality of meat and lack of waste in bone and loose fat, were all carried to the highest points of perfection in the Aberdeen-Angus.

In this connection, it might be well to mention here that the world's dressing percentage for a beef animal is held by an Aberdeen-Angus heifer. Luxury, Smithfield champion in 1885, dressed 76¾% at 2 years and 8 months of age. The Michigan Agricultural College, Lansing, Michigan, also reports an Aberdeen-Angus cross that dressed over 73%.

Hitting the commercial cattle market and thus demonstrating the practical points of Aberdeen-Angus pedigrees has always been the aim of the men behind the breed in America. Mr. W. C. McGavock of Illinois is authority for the statement that Aberdeen-Angus for fourteen successive years prior to 1903 furnished the highest priced carload of cattle sold on the American markets. With Carroll County, Mo., as a breeding-feeding center built up around the herd of the Reas, who sent the first consignment of cattle to American markets carrying Aberdeen-Angus blood in 1880, his neighbors quickly learned of their excellent feeding qualities and fed out market toppers bred in this section. R. B. Hudson in 1889 obtained $7.10 for the highest priced carlot of the year. Thos. Brandon received $7.40 as the highest price of 1890; W. C. White, in 1891, received $7.15; all of these men were from Carrollton. J. D. Eubank, Slater, Mo., was high man in 1892 at Chicago, selling the only load to bring $7.00 that year. In 1893, W. C. White again had the year's top, getting $7.00. In 1895, J. Evans, Jr., & Son, Emerson, Iowa, had the market toppers of the year at $6.60. In 1895, W. C. White, of Missouri, repeated earlier honors at the market by topping for the year at $5.50. L. H. Kerrick, a brother-in-law of the Funks of Bloomington, Ill., and a stalwart in the Aberdeen-Angus ranks, took the honors two years in succession by selling loads at $5.90 and $6.00 for 1896 and 1897, respectively. Mr. Kerrick again repeated top price honors in 1898 and 1900 with $8.25 and $15.00 respectively and J. Evans, Jr., & Son, also prominent breeders as well as feeders, were top in 1899 at $6.25. These were all Chicago prices.

In 1901, however, at the Pittsburgh fat stock show, Chas. Escher, Sr., showed the champion load which sold for the high price of the year, $21.50, a record for carlot sale on any market until the war period. Mr. Escher repeated top selling honors on a carlot in 1902 when his load was champion at the International Live Stock Exposition, selling at $14.50.

Claus Krambeck of Iowa fed the International grand champions of 1904, 1905 and 1907, and while he has passed on, his sons still breed and feed the kind their father taught them to win with at Chicago. The Funk Brothers, Bloomington, Ill., of seed corn fame, and brothers-in-law of Mr. Herrick, showed the grand champion loads of 1906 and 1908. Practical cattle feeding of cattle bred by themselves and neighbors was developed to a high point by this firm, who saw in breeding only a more direct route to turning out a more economical product for the market.

In 1910, there appeared in the carlot division of the International a younger cattle feeder whose keen cattle sense was backed by generations of cattle feeders, and the grand championship fell to him, the first of five such honors inside the decade to follow. Mr. Hall's father and grandfathers had all been practical cattlemen who had seen the cattle business develop from the days before the railroads, packing houses and refrigeration. They had been drovers and feeders as well, feeding the cattle gathered up in their territories of Kentucky and later central Illinois, and driving them to New York, Boston, Philadelphia and Baltimore after fattening. Mr. Hall knew all there was to know about the producing end of the commercial cattle business, and he carefully studied out the consumers' end by watching the practical packer buyers select the grand champion carlots at the earlier Internationals. Then he set out to gather the kind of cattle that could be fed out to win this honor for himself. That study brought him to the Aberdeen-Angus and the high honors of 1910, 1912, 1916, 1917 and 1920. Escher & Ryan, the famous breeding firm of Iowa, furnished the grand champions and reserve grand champion carlots of 1911 and 1913. In 1919 and in 1921, John Hubly, an Illinois feeder, won the honors on carlots with Colorado-bred yearlings.

Thus, it will be seen that following the 14-year-top run up to 1903, pointed out by Mr. McGavock, there has been an 18-year-top run with two breaks for the Shorthorn load named in 1909 and the Hereford load of 1918. In other words, for 33 years, with three exceptions, Aberdeen-Angus have sold for the annual record price for carlots. Practically half of these record-priced loads have been the grand champion fat carlots of the International, but those grand champions were picked by practical men of the stock yards commission companies and the packing houses as the best loads of the years they showed.

AMERICAN BREEDERS ORGANIZE

The American Aberdeen-Angus Breeders' Association was incorporated in 1883, under the laws of the State of Illinois—just ten years after the first importation by Mr. George Grant, Victoria, Kansas. The petition for the charter of the Association was signed by William T. Holt, Charles Gudgell, H. W. Elliott and A. B. Matthews. The directors for the first year were: W. T. Holt, Denver, Colo.; Col. John Geary, London, Ont., Canada; H. C. Burleigh, Vassalboro, Me.; Charles Gudgell, Independence, Mo.; Stephen Peary, Trenton, Mo.; Wallace Estill, Estill, Mo.; A. M. Fletcher, Indianapolis, Ind.; and Abner Graves, Dow City, Iowa, who now lives in Colorado and to whom we must give credit for the breed name, "Aberdeen-Angus." W. T. Holt was elected first president, Mr. Burleigh, first vice-president, and Charles Gudgell, secretary and treasurer. In 1888, Mr. Gudgell was succeeded in the secretaryship by Thomas McFarlane, Iowa City, Iowa, and two years later the offices of the Association were removed from Iowa City to Harvey, Ill., a manufacturing suburb of Chicago, founded by T. W. Harvey, the laird of Turlington. In 1902, the headquarters of the Association were located in the Pedigree Record Building, Union Stock Yards, Chicago, the most natural place for it. In 1907, Mr. Chas. Gray became secretary.

Following W. T. Holt, the presidents have been: H. C. Burleigh, Vassalboro, Me.; A. M. Fletcher, Indianapolis, Ind.; George Geary, London, Ont.; T. W. Harvey, Chicago; W. A. McHenry, Denison, Iowa; H. W. Elliott, Estill, Mo.; E. S. Burwell, Madison, Wis.; M. D. Evans, Emerson, Iowa; O. E. Bradfute, Cedarville, Ohio; L. McWhorter, Aledo, Ill.; W. F. Dickenson, Redwood Falls, Minn.; L. H. Kerrick, Bloomington, Ill.; E. T. Davis, Iowa City, Iowa; J. S. Goodwin, Chicago; George Stevenson, Jr., Waterville, Kan.; W. J. Miller, Newton, Iowa; M. A. Judy, West Lebanon, Ind.; C. E. Marvin, Paynes Depot, Ky.; A. C. Binnie, Alta, Iowa; Stanley R. Pierce, Creston, Ill.; H. J. Hess, Waterloo, Iowa; John D. Evans, Sugar Grove, Ill.; E. F. Caldwell, Burlington Junction, Mo.; H. M. Brown, Hillsboro, Ohio; Chas. Escher, Jr., Botna, Iowa; J. Garrett Tolan, Farmingdale, Ill.; J. C. White, Winterset, Iowa; O. V. Battles, Yakima, Wash.; L. A. Campbell, Utica, Minn.; Carl A. Rosenfeld, Kelley, Iowa.

In 1888, membership, which had been originally fixed at 200, was made unlimited, and the number at present is over 6,000. In the various volumes issued by the Association there has been entered a total of 350,000 animals, each volume now receiving successive accretions of about 18,000. In 1890, $3,000 was set aside for the World's Fair Fund, 1893. The membership fee was raised from $10 to $20.

DEVELOPMENT OF ASSOCIATION ACTIVITIES

The progress of the breed has not all been plain sailing without any direction or "pushing" in North America. Show prizes have always been a means of putting the breed before the farmers, feeders and ranchers, and large amounts have always been appropriated by the Board of Directors of the American Aberdeen-Angus Breeders' Association. With founding of the International Live Stock Exposition, annual appropriations of $5,000 were put up for this great show alone, and at that time, this was a heavy drain on the young Association. For 1920, show, fair and calf club appropriations totaled $40,000.

RISE OF PUBLIC SALES

The development of the public sale business shows conclusively the way the breed is going ahead in the United States, seeding down the territory where it has got a start, and spreading all over the fields wherever new seed falls from the herds already started. In 1919 the high tide was reached when 114 sales distributed 5,412 head of pure-breds at an average of $511.59, or a total business of $2,768,761.50. P. J. Donohoe & Sons, Holbrook, Iowa, sold 46 lots for an average of $2,626.85 on May 21, 1919. Escher & Ryan, at Manning, Iowa, sold Enlate for $36,000 on June 3, a new record for a bull of the breed at public auction, and the next day they sold Blackcap McHenry 131st for $10,000, a new female record for the Doddies; and their sale of 174 head totaled $373,650.00, an average of $2,147.44.

Still higher was the "war peak" of 1920, when over 5,000 head sold at an average of nearly $706 in more than a hundred sales. Record average for a single sale, record price for a female, as well as record average for any beef breed for a year all went to the Aberdeen-Angus during this year of after-war inflation just before the deflation period. New territory in the South was opened up, as well as old territory enlarged and rejuvenated. Sales in Tennessee went above $1,000 average, and that state became one of the leaders with the breed.

The McHenry herd, at Denison, Iowa, became America's leader and the constructive breeding work done there put McHenry in the position enjoyed by Sir George Macpherson Grant of Ballindalloch fame in the old country. The sale at private treaty of the herd in 1914 to Escher & Ryan made that Iowa firm easily the leader in America. Built up as it had been with the best of Ballindalloch blood from several importations, notably one that brought the Perth champion, Earl Eric of Ballindalloch, to this country, it became irresistible in show and sale ring and wrote history in large letters from the date of the addition of the McHenry herd down to the present day. The Escher importation of Trojan Erica heifers in 1909 injected a large stream of Scotland's most fashionable blood into a healthy stream already flowing through American feed lots and show rings.

In 1916, Clarence W. Eckardt of New York City imported Evenest of Bleaton and a little group of heifers as a foundation herd. The naming of Evenest of Bleaton as grand champion at the 1918 International Live Stock Exposition, did much to direct attention to the breed in the East at a time of influx in American agriculture.

The importation of 120 head by the famous Shorthorn firm of Carpenter & Ross for a sale at Chicago and a foundation herd in Ohio, in 1920, was another milestone in the history of the breed in America. Campbell Brothers of Minnesota also imported a herd bull, while Woodcote Farm of Michigan brought in two herd bulls and some choice females. The importation, in 1918, of Edgar of Dalmeny by Wildwood Farms, also of Michigan, attracted attention.

IN CANADA

In 1876, Professor Brown, an Aberdonian, who occupied at that time the position of director of the Ontario Experimental Farm at Guelph, imported the bull Gladiolus and the heifers Eyebright and Leochel Lass 4th.

Prof. Brown imported a second group of five head, two bulls and three cows, that was destined to have a far-reaching effect on the breed in Canada, and the northern states of the United States. The bull, Strathglass cost the Guelph station $2,500, a rare price in those days. The 4-year-old cow, Kyma, with Strathglass can be found in practically all of the best herds of Canada today, and one International Live Stock Exposition carcass carried a very heavy infusion of this blood from the Elm Park herd of James Bowman. A bull calf from Advie Mains, and two other cows from Mr. Bennett's dispersion sale made up the five head bought for the Ontario Station in this importation. These formed the nucleus of the breed in Canada, achieving a reputation not only as beef cattle, but as producers of milk rich in butter fat.

Hon. M. H. Cochrane founded one of the most valuable herds in this hemisphere by purchases first made in 1881, from Glamis, obtaining Beauty of Glamis (3515), an Erica-topped specimen of that family. He also secured the finest specimens from Powrie, Waterside, Corskie, Easter Skene, Guisachan, Altyre and other herds. Blackbird of Corskie 2d (3024), the first-prize cow at Perth, 1879, by John Bright, representing the Montbletton Mayflower family, was one of the cracks of the herd. Mabel 6th (4295), a Pride, was got from Methlick; Vine 2d from the Earl of Southesk and Pride of Aberdeen 20th from Bridgend. Most of the Scotch herds were drawn on, twenty-five bulls being selected at one time for the Cochrane Ranch Company, in the Northwest Territory.

Mossom Boyd, Bobcaygeon, Ont., established a herd in 1881, founded upon old family material. The herd made a marked impression in Ontario and was one of the best ever put together. During its existence it had a splendid record at the Provincial and other shows.

Hon. J. H. Pope, Dominion Minister of Agriculture, also formed a herd in 1881, the number purchased being fifteen heifers and a bull. Included in the selection were Charmers (Queen Mothers), Zaras, Castle Fraser Minas, Ballindalloch Lady Fannys and other specimens of the Queen Mother tribe.

The Model Farm herd of George Whitfield, Rougemont, Quebec, was selected by John Grant, Bogs of Advie, among the lot being Judge, the Ballindalloch exhibit at Paris in 1878, that later went to the Heatherton herd. This selection was followed by a second, consisting of still higher-bred specimens, including Ericas, Queen Mothers, Jilts, Montbletton Mayflowers, Rothiemay Georginas, Drumin Lucys and Westertown Roses.

The Hudson's Bay Company took a considerable bunch of imported Aberdeen-Angus into the Northwest Territory in the 80's, and a considerable part of the herd went to found the Glencarnock herd of the McGregors at Brandon, Manitoba, later. Some bulls from the Canadian East, and a strong importation from Scotland about 1909, with strong infusions in Canada. Its two grand champion steers at 1912 and 1913 International shows, with grand champion bull in 1912, indicates the strength of this herd. Howard Fraleigh, John Lowe, James Bowman, G. C. Channon, Robert McEwen, and J. D. Larkin are at present some of the leaders in eastern Canada. In the West, herds are building up to strength with the growth of the country. G. N. Buffum, Browne Bros. in Saskatchewan; Chas. Ellett, A. E. Noad, S. Henderson and G. C. Montgomery in Alberta and E. C. Harte, James Turner, John R. Hume are some others in Manitoba.

IN THE DUAL PURPOSE FIELD

The record of the breed as a beef champion has become so thoroughly established that it might be supposed it has made no pretensions to milking honors. Nevertheless, the Aberdeen-Angus breed produced the champion at the great show of the British Dairy-Farmers' Association, held in London, 1892.

The victory was certainly a great feather in the cap of the breed and the Aberdeen "Free Press," in reporting the event, said: "To those who know the history of the breed, the position taken by J. F. Spencer's six-year-old polled cow, Black Bess, will hardly occasion surprise. The victory will probably stimulate breeders to give more attention to the milking qualities of their cattle." This would indicate that good milkers were common enough then to occasion no great remark.

The number of points on which this victory was won, was the highest ever scored at that show and the cow was described as a quite first-rate specimen for the purpose of town milk-sellers—"The special clients of the British Dairy-Farmers' Association."

This cow was not a singular instance, by any means; there are many like her being bred today. What is inherent in a breed can by training be discovered and developed, and that heavy milking quality is inherent in the Aberdeen-Angus breed is very easily proved.

From the Scottish Farmer's Magazine we were able to glean the following records of the oldest herd now in existence, that of Robert Walker of Portlethen, referring to the year 1845: Cow No. 1, seven years old, had twin calves three times, and has been but once dry since she calved in 1839, 3,024 pints, Scotch, or about 7,500 lbs.; No. 2, eight years old, 2,931 pints (7,236 lbs.); No. 3, seven years old, 2,388 pints (6,000 lbs.), twice had twins; No. 4, seven years old, 2,020 pints (5,000 lbs.), has had seven calves. Another, eleven years old, gave 1,830 pints (4,575 lbs.). The last was a prize cow at the national shows, as likely were some of the others, for Robert Walker was a highly successful exhibitor in the regular classes of the breed at the national shows, and these records justify the idea of the breed being a dual-purpose kind.

Sixteen to eighteen and even twenty Scotch pints (40 to 50 lbs.) were not uncommon daily records noted among the progenitresses of the leading tribes, and they were "steady milkers" all the year round. In later years Mr. Fullerton, having to stock his farm with other cattle than the native polled, on account of having suffered losses on three occasions from rinderpest and pleuropneumonia, was wont to dilate upon the milking qualities of the early polls and "their ready tendency to fatten and also milk well."

Old Lady Ann (743) of the old Kinnaird Castle herd, believed by Chas. Carnegie to have been the oldest cow entered in the Scotch herd book, had a host of descendants, all excellent milkers, having the especial faculty of continuing to give a large quantity of milk till close on the time of calving, and, if allowed, would continue to milk until they did calve. From those who observed the milking propensity of Old Lady Ann and her descendants, it was believed that many of them gave more milk than any of the Ayrshires, from one year's end to another; though possibly not so much immediately after calving. Lady Carnegie used to recount the traditions of the milking qualities of the old Kinnaird polls—they were hereditary milkers. Emily, the dam of Erica, was a most valuable dairy cow, like Black Meg.

Mary Ann of Ranniston, that sold for the highest price at the Bridgend sale in 1857, was described as "a very deep milker;" Young Kate, bred by Alexander Bowie, was described as "the best milker" in the Mains of Kelly herd. Rosalie of Bogarrow (27198) gave 45 lbs. a day for a month after calving. George Dickenson, one of the first to establish a herd of Aberdeen-Angus in England, has put it on record that certain strains of the breed gave fully the average quantity of milk given by any breed, the quality being second only to that of the Jersey. But many assert that the quality surpasses even that of the famed island breed.

William Robertson, Aberlour, Mains, always gave much attention to the milking properties of his herd; and his experience was that, "by very little extra trouble, it was possible, without sacrificing the merits of the breed in the matter of beef production to rear animals that would yield a copious supply of milk of choice quality."

Lord Airlie, one of the substantial patrons of the breed, paid particular attention to the dairy qualities of his favorites. From 35 to 40 lbs. was a common daily record in his herd. His cow, Bell of Airlie, used to milk all the year round. The Drummuir herd was also a striking example of how well the polls responded to the pail test. Nothing was kept for breeding which did not show promise in this direction, and the herd, developed along such lines, simply proved the early claims made for the breed. The Castle Craig herd, too, was founded by Sir Thomas D. Gibson-Carmichael upon milking lines, all the original selections being made for the general utility, especially as milk producers.

The practical value of the Aberdeen-Angus milking propensity has, however, long been familiar to discerning American breeders. J. H. Moore, Cook County, Ill., maintained a herd of pedigreed and grade Aberdeen-Angus cows as a working, money-making dairy. His experience led him to the conculsion that well selected grade Aberdeen-Angus cows are valuable for dairy purposes just as much as they are for beef, and that the quality of their milk and the continuance of the period during which they may be relied upon to yield a supply are both much in their favor. The Oatman Condensed Milk Company, to whom Mr. Moore consigned the milk reported a test made at their factory of Mr. Moore's registered and high-grade Aberdeen-Angus cows. The samples tested on November 22, averaged 5.85 per cent butter fat, and these tested a week later averaged 5.32 per cent. The test made from the milk of the entire herd of grade Aberdeen-Angus cows averaged from 5 to 4.50 per cent butter fat for the entire season, which was one of the highest, if not the highest, test of milk from any dairy coming to the factory.

Prof. Brown of the Ontario College of Agriculture, made extensive tests in regard to the specific gravity of milk from different breeds and found that the Abedreen-Angus breed recorded 111.0; the Hereford grade, 106.0; Shorthorn grade and the Ayrshire, 103.0; Hereford, 91.0; Shorthorn, 86.0. When the records for yield of butter from milk by weight were secured they showed that the Aberdeen-Angus also stood first with 3.72 per cent, followed by Hereford grades 2.54 per cent. Shorthorn grades, 2.31 per cent, and Herefords, 2.01 per cent.

In a series of tests made in the county of Banff, Scotland, in which the best stocks of the county were represented, Mr. Findlay of Aberlour supplied cream from the cows of his herd which produced from seven quarts a return of nine pounds of butter; while the Countess of Seafield's herd supplied eight quarts of cream producing ten pounds of butter.

A very striking demonstration of the milking qualities of the breed is supplied by the determination of John Moir, in far away Australia, to maintain the reputation of the breed as to milking qualities, against all disparagement. Mr. Moir issued a challenge to the breeders of all cattle in South Australia, and as a result of the test the Aberdeen-Angus cattle came out easily victorious, Mr. Moir showing two cows both over twelve years old, that were giving twenty quarts a day, and had been doing so for three months after calving. He also cited the case of a herd of polls in the coldest district of Victoria that milked satisfactorily through the exceptionally severe winter, while the other breeds were dying off by the score of exposure and starvation. These are only several instances mentioned by Mr. Moir, who thus scored for the breed a decisive triumph on that continent. The result was gratifying; though owing to the scarcity of heifers of the breed, the cattlemen went ahead raising calves from Ayrshire cows and polled sires; and so there arose a keen demand for Aberdeen-Angus cattle both for beef and for dairy purposes. The leading Sydney milk purveyor, F. A. MacKenzie of Waverly, mentioned a cow, Emily, a poll cross, winner of the championship at the Sydney Royal in 1901, which had for the prior three seasons been giving close to 60 lbs. of milk a day for almost three months after calving. His champion cow of the year following at Sydney, also a black polled cross, gave at the show tests 134.14 lbs. of milk and 6.95 lbs. of butter in three days. "The quality, texture and flavor of the butter were of the very best, and freely commented on as such." Many other facts of the same kind come to us from Australia, Mr. Moir supplying examples that are convincing. He states that Mr. Kerr of Glenroy, near Melbourne, who milks 500 cows a day, assured him that some of the best cows he ever milked were Aberdeen-Angus, further adding that he had never seen one with sore teats. Mr. Beaty, Toolern, Victoria, with some twenty cows of the breed had five giving over 20 quarts a day, and one 23½ quarts—the poorest giving 14 quarts.

Again, later in Canada, we find Mr. Bridges of Surrey, England, reporting the case of a farmer who bred Aberdeen-Angus bulls to Ayrshire cows, the crosses obtained being the best milkers he ever possessed. In Argentina, dairy farmers in the vicinity of Bahia Blanca find an excellent cross for dairy purposes is obtained by using Aberdeen-Angus bulls on their Holstein cows.

It is a very well known fact that the old Buchan Polls were famous in the dairy way—equal in the opinion of many who compared them to the Ayrshire breed. In 1805 the total products of the dairy in the county of Aberdeen was valued at $1,150,000, which was just $100,000 less than the value of the total number of cattle killed or exported for beef, and the principal amount of which was derived from the Buchan district. The Buchan cows had such a dairy fame that Sir John Sinclair, the founder of the Board of Agriculture and the developer of the famous Statistical Account of Scotland, and of the "General Views of the Agriculture of the Counties" who was therefore well posted upon the best everywhere in the United Kingdom, sent into Buchan for specimens to go into Caithness, his own county, "to fill the dairy."

James Henderson, writing in 1826, said: "In the district of Buchan the cows are much famed for the Dairy." It is also interesting to know that Sir John Macpherson Grant, grandfather of the present baronet, we presume, also sent to Buchan for specimens of the breed, these being among the first, if not the first, polled cattle we can trace as existing at that headquarters of the polled.

The standard work on dairying published in Britain is Professor Sheldon's "Dairy Farming," the first edition of which appeared in 1879, and in it we find some very pertinent remarks relating to the milking qualities of the Aberdeen-Angus breed—which, first referring to the general idea of the public as to the lack of milking qualities in the breed, continue thus: "There are, however, many excellent milkers among the Aberdeens, and it is safe to assume from these instances that the breed is not by any means destitute of the qualities which go to make up a first-class dairy breed. The one thing needful is to cultivate, as has been done in other breeds, the development of the lacteal organs. Were they treated as the Ayrshires or the Shorthorns have been, there can be little doubt of their improving in milking properties.

Dr. Thos. F. Jamieson, Lecturer on Agriculture in Aberdeen University, in a lecture remarked on the neglect of the former milking qualities of the breed and added: "Valuable prizes should also be given in the proper classes for animals uniting fine symmetry with good dairy qualifications. If the breeders would also afford some information in their printed catalogues as to the milking pedigrees of animals they offer for sale, I think they would soon find that their customers would appreciate it, and that animals well come in this respect would be looked after, and would fetch high prices at their sales."

The breed's whole history proves that basically it was always what is now called a dual-purpose breed. The breeders not only bred for beef, but for milk. They had rules set down which they had followed for generations previous in 1813, the date of Skene-Kieth's famous "View." These rules were:

"1. For beef—the animal should be handsome, well-formed, short-legged, with a smart or keen eye, and a rough ear.

"2. For milk—a small neck and head, broad in the hind quarter, her bag or udder lying well forward on her belly, and her teats, well spread, or pretty distant from each other."

LITERATURE ON THE BREED

The following embraces what may be considered the standard British and American works and articles on the breed:

"Cattle and Cattle Breeders," by James McCombie, 1867. Third edition revised by James Macdonald, Secretary of the Highland and Agricultural Society, 1882. (Blackwood & Sons, Edinburgh.)

"History of Aberdeen-Angus Cattle," by James Macdonald and James Sinclair, 1882. Second edition, 1910, revised by James Sinclair, Editor of the Livestock Journal, London. (Vinton & Co., London.)

"The Breed That Beats the d, a Demonstration of the Properties, Prepotence, Pre-eminence and Prestige of Abero Angus Cattle," 1884, by R. C. Auld. (Aldine Co., Detroit.) Out of print.

"Aberdeen-Angus Cattle, Being Notes on Fashion and an Account of the Leading Families of the Breed," by Albert Pulling. (Simpkin, Marshal, Hamilton, Kent & Co., London, and William Pile, Ltd., 26-High-St., Sutton.)

"Aberdeen-Angus Cattle, Their Recent History," by Jas. R. Barclay. ("Banffshire Journal," Banff.)

"Famous Aberdeen-Angus Sires," by George Henry. (Transactions of the Highland and Agricultural Society, 1898, fifth series, Vol. 10.)

"A History of the Ballindalloch Herd of Aberdeen-Angus Cattle," by C. Macpherson Grant. ("Banffshire Journal," Banff.)

"Aberdeen-Angus Cattle on the Range," the classic exposition of the breed as Range Cattle by George Findlay. (Breeder's Gazette, 1899.)

"Aberdeen-Angus Cattle and Their Crosses as Beef Producers," by Albert Pulling, Secretary, English Aberdeen-Angus Association.

"Report of Kansas State Board of Agriculture on Polled Cattle, 1897," by F. D. Coburn, Secretary, Topeka, Kansas.

"Aberdeen-Angus Cattle, Breeding and Management," by Clement Stephenson, D. Sc., F. R. C. V. S. Proceedings of Armstrong College Agricultural Student's Association, 1908-9.

"Management of Aberdeen-Angus Cattle," by Clement Stephenson, D. Sc., F. R. C. V. S. (Journal of Royal Agricultural Society of England, 1894, Volume V. Third series, Part 1.)

"Supremacy of Aberdeen-Angus Cattle." Results of leading fat stock shows in Great Britain and America. (Edited by Chas. Gray, Secretary American Aberdeen-Angus Breeders' Association, 817 Exchange Ave., Chicago.)

STANDARD WORKS HAVING CHAPTERS ON ABERDEEN-ANGUS

There are also articles on Aberdeen-Angus cattle in the following standard works:

Stephens' "Book of the Farm"; revised edition by James Macdonald. (W. Blackwood & Sons, Edinburgh.)

"The Complete Grazier"; revised edition by Dr. W. Fream and W. E. Bear. (Crosby, Lockwood & Son, London.)

Prof. Wallace's "Farm Live Stock of Great Britain." (Crosby, Lockwood & Son, London.)

"Encyclopedia of Agriculture"; edited by C. C. Green and David Young. (William Green & Sons, Edinburgh.)

"Standard Cyclopedia of Modern Agriculture"; edited by Prof. R. Patrick Wright. (Gresham Publishing Co., London.)

Live Stock Handbooks—"Cattle—Breeds and Management." (Vinton & Co., Ltd., London.)

"Chronicles of Aberdeen-Angus Cattle," serially in Breeder's Gazette, 1887, by R. C. Auld.

The Farming News Annal, Studs and Herds number, Perth, Scotland, also contains an annual review.

Farmers' Bulletin 612, "Breeds of Beef Cattle," U. S. Department of Agriculture, Washington, D. C.

"Aberdeen-Angus Cattle" in Bailey's Encyclopedia of American Agriculture. (Macmillan Co., New York, 1905.)

"The Live Stock Journal Almanac" contains an annual review of the breed happenings, by Jas. R. Barclay, of the Aberdeen-Angus Cattle Society, formerly by Geo. Hendry.

The "Aberdeen-Angus Review," published semi-annually by the Aberdeen-Angus Cattle Society, Aberdeen, Scotland.

Back numbers and bound volumes of the Breeder's Gazette for show ring, sales and special articles and news.

The Aberdeen-Angus Journal, Webster City, Iowa.

The Canadian Aberdeen-Angus Recorder, Brandon, Manitoba, Canada.

www.ingramcontent.com/pod-product-compliance
Lightning Source LLC
Chambersburg PA
CBHW081122240526
45470CB00019B/2919